D1574482

Contemporary Problems in Geography

The general editor of *Contemporary Problems in Geography* is Dr. William Birch, who is Director of the Bristol Polytechnic. He was formerly on the staff of the University of Bristol and the Graduate School of Geography at Clark University in the USA and he has been Chairman of the Department of Geography in the University of Toronto and Professor of Geography at the University of Leeds. He was President of the Institute of British Geographers for 1976.

Alan Wilson is Professor of Regional and Urban Geography at the University of Leeds. After reading mathematics at Cambridge he has served as Scientific Officer at the National Institute of Research in Nuclear Science, Research Officer at the Institute of Economics and Statistics, University of Oxford, Mathematical Adviser at the Ministry of Transport and Assistant Director of the Centre for Environmental Studies. His publications include *Entropy in Urban and Regional Modelling, Papers in Urban and Regional Analysis* and *Urban and Regional Models in Geography and Planning.*

Michael Kirkby is Professor of Physical Geography at the University of Leeds. He has done research at Cambridge and at the Johns Hopkins University and previously taught at the University of Bristol. In 1972 he published *Hillslope Form and Process* with M. A. Carson.

Systems Analysis in Geography

RICHARD HUGGETT

CLARENDON PRESS · OXFORD
1980

Oxford University Press, Walton Street, Oxford OX2 6DP

OXFORD LONDON GLASGOW
NEW YORK TORONTO MELBOURNE WELLINGTON
KUALA LUMPUR SINGAPORE JAKARTA HONG KONG TOKYO
DELHI BOMBAY CALCUTTA MADRAS KARACHI
NAIROBI DAR ES SALAAM CAPE TOWN

Published in the United States by
Oxford University Press, New York.

British Library Cataloguing in Publication Data

Huggett, Richard
Systems analysis in geography.
(Contemporary problems in geography).
1. Geography–Methodology
2. System analysis
I. Title II. Series
910 G70 79-41097

ISBN 0-19-874081-6

ISBN 0-19-874082-4 Pbk

Filmset by University Press, Belfast
and printed in Great Britain by
Richard Clay (The Chaucer Press), Ltd,
Bungay, Suffolk

Preface

Confront a geographer with the phrase 'systems analysis' and he may inwardly laugh, cry, or regard in awe: some regard it as an even better geographical joke than the *Guardian's* April Fools' Day advert for that small and balmy country called San Seriffe; some despair at the thought of it; others see it as a passport to a geographical paradise in which all problems are solved by a liberal dose of systems theory. The aim of this little book is at once to show how systems analysis does help to solve some geographical problems and to allay fears as to the difficulty of systems analytical methods. In doing so, the philosophical aspects of systems methods are not enlarged upon because an in depth study of these is not required in evaluating systems analysis as a purely investigative tool in the geographer's kit-bag.

Written as an introduction to systems analysis for undergraduates, emphasis is given to illustrative examples rather than difficult mathematical expositions. Nevertheless, maths is the workaday language of systems analysis and a certain amount of it is unavoidable; where it is used, it should not be beyond the grasp of students who are taking introductory techniques courses and nowhere takes the dazzling flights of formulation found in some texts: all the examples can be understood without reference to the formulae and so the mathematical novice need not fear.

I have a list of people to whom I wish to express my thanks: for fostering my initial interest in geography, H. W. (Masher) Martin; for friendly tutoring and scintillating discussion during my under-graduate and post-graduate years, Andrew Warren and Claudio Vita-Finzi; for drawing the diagrams, Clive Thomas; for helpful comments on the first draft, Professor Alan Wilson; and finally, for forbearance during the gestation period of the book, my family.

Macclesfield,
November 1978

RICHARD HUGGETT

Contents

1 What is a System?

Every work, both of nature and art, is a system.
BUTLER, *The Analogy of Religion*

1.1 System structure

WHAT are the systems studied by geographers and how are they defined? Let us start by defining a system. Broadly defined, a system is a set of interrelated parts. Other definitions abound. A favourite one among geographers is that of Hall and Fagan (1956) which runs: '... a set of elements together with relations between the elements and among their states'. Here is an even wordier one from Miller (cited by Laszlo, 1972): '... a natural system is a non-random arrangement of matter-energy in a region of physical space-time, which is non-randomly organized into co-acting interrelated subsystems or components'.

It can be gleaned from these definitions that a system has at least three basic ingredients—elements, states, and relations between elements or states. The elements or components of a system are either physical objects—clay particles, people, towns, and so on; or concepts—words, numbers, and the like. According to Ackoff (1971), physical objects are components of concrete systems whereas concepts are elements of abstract systems. This distinction is important in the geographical realm since geographers deal with both the world of objects and the world of the mind—the real world and the perceived world. In this book we shall be mainly concerned with concrete (physical) systems and, unless otherwise qualified, this is what the term system will mean. Each system element has a set of properties or states—number, size, mass, colour, age, price, and so forth, each of which is termed a state variable. The system state is defined by the values of the state variables at a particular point in time. So the state of an urban system may be given by listing the values of variables such as population, employment, and land availability at a particular time.

The restriction of the system to a set of elements implies that systems are bounded. State variables outside the system, or the environment as it is known, are exogenous variables, those inside the system are endogenous variables. In a drainage basin, where the boundaries are the watershed and the land surface, precipitation is an exogenous variable whereas soil wetness and streamflow are endogenous variables. Systems which lack exogenous variables are said to be unforced systems, those with exogenous variables are forced systems. Most systems in geography are forced ones. Without energy and material supplied to them by exogenous variables, most systems would run down. Because of this, exogenous variables are sometimes referred to as forcing functions or driving functions. The major forcing functions in a drainage basin are precipitation and incoming radiation. For short periods of time, however, a drainage basin could be studied as an unforced system; for example, when looking at water redistribution during a prolonged dry period. A distinction is often made between systems whose boundaries are open to the transference of matter and energy—open systems, and systems whose boundaries are open to the passage of energy only—closed systems.

Relations between state variables or between system elements may be expressed verbally, as say cow eats grass; statistically, as say a correlation between milk production of a cow and the amount of grass consumed; and mathematically, as say a function relating milk production to grass consumption. System relations do not arise by chance: a system is not a mere aggregate. In a general way anything consisting of distinguishable elements could be termed an aggregate; but it is preferable to restrict the term aggregate to component parts which are loosely or accidentally associated such as sand grains in a heap of sand. In a system, relations between components are more intimate and the entire assemblage is adjusted as a regular and connected whole.

Systems are normally defined in terms of three levels: the level of interest, or the system inself; a level within, or the system components; and a level without, or the environment (Simon, 1962). Thus, in geography, the system of interest might be a regional communications network which is closely related to links between central places; the level within would be the social communications network which links individual persons; and the environment would consist of national and international communications networks which are shaped by metropolitan fields, political allegiances, and so forth (Gould, 1969; Hägerstrand, 1967) (Figure 1.1). Likewise, a community of people may be the system of interests; a community is divisible into families and individuals which form the level within; and, at the same time, a community is an element in a system of communities, such as a city, which is the level without, or environment. Many systems of interest in geography have such a hierarchical organization. Systems at each level of a hierarchy display a degree of autonomy over their component parts and over their environment. The autonomy is achieved in part by the system at a given level functioning at a different space and time scale from systems at other hierarchical levels. Thus in the regional communications network, which we shall take as the level of interest, the relatively fast and frequent contacts between individuals in the micro-scale, social-communications network pass virtually unnoticed

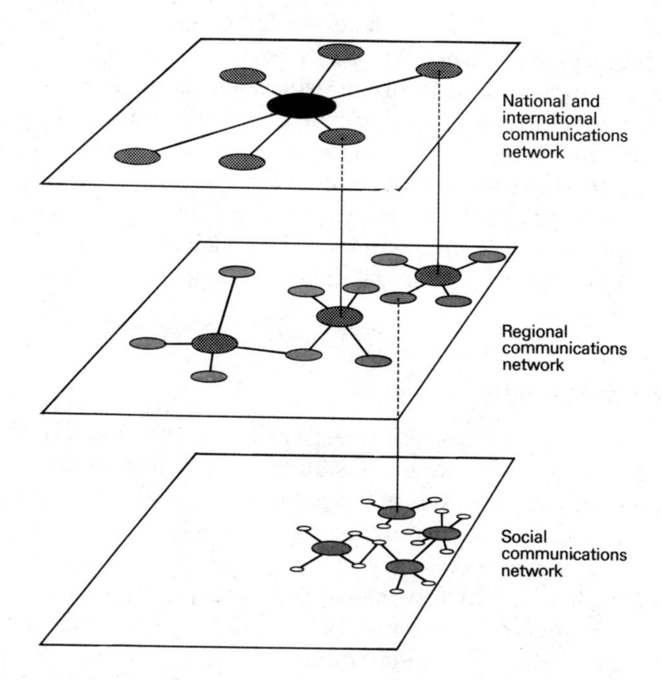

FIG. 1.1 A system hierarchy: the national, regional, and local social communications networks. Based on Hägerstrand (1967).

at the regional level; instead, they appear to be in a steady state. At the other extreme, communications at a national and international level are relatively slow and infrequent compared with those at the regional level where they appear as roughly constant influences on the regional system's behaviour—they are exogenous or driving functions which are seen as regular and well behaved. Thus the dynamics of the system of interest, the regional communications network, act virtually independently of the fast dynamics of the inter-regional social communications network, and the slow dynamics of the national and international networks.

Despite the almost autonomic behaviour of a system at any level in a hierarchy, there is a degree of relationship between the system of interest, its parts, and its environment. The relations are bipartite. On the one hand, the system is linked to its component sub-systems: on the other hand the system is linked to a larger, super-system, or environment, of which the system is but a component part. A system at any level of interest thus has what Koestler (1969) called two faces: the face turned towards a lower level, towards the system parts or sub-systems, is that of an autonomous whole; the face turned up, towards the environment or super-system, is that of an independent part. Koestler (1967) coined the term 'holon' (from the Greek *holos* = whole, plus the suffix *-on*, as in proton, suggesting a particle) to describe these Janus-faced structures. Smailes (1971) introduced to urban geography the idea of the holon: he suggested that the ranks in the urban hierarchy (cities, towns, villages, and so on) are holons in so far as they display some properties of autonomous wholes (they may have their own administration for instance) but also possess properties of dependent parts (their administration will normally run within constraints imposed by higher-level administration). In this book, the term 'system' will be retained in favour of its very close relative, the 'holon'.

1.2. System dynamics

Most geographical systems of interest are dynamical: their state changes with time. Changes in a system's state can be traced within a system's phase space, that is, the space produced using the values of state variables as mutually orthogonal co-ordinates. Take the instance of two populations, one of which preys upon the other; this system has a two-dimensional phase space (or phase plane) created by plotting one state variable—the size of the prey population, against the other state variable—the size of the predator population. In the general case of a system with n state variables, there will be an n-dimensional phase space defined by n carterian co-ordinates, one for each state variable. Any point within the phase space uniquely defines a set of values for the state variables, in the example the size of the prey and predator populations, at a particular point in time. If the state of the system should change with time, a path or trajectory will be traced through the phase space, the direction taken depending upon the relations between system components. In simple cases the system state may do one of three things. Firstly, system state may move towards a point or node in the phase space which is independent of the initial state of the system, and in which state the system tends to remain—this is the steady state and in this condition the system is said to be stable (Figure 1.2a). In the predation example, this means that, owing to the predator's inefficient catching of its prey, a balance between prey and predator population numbers will exist. In fact, the relations between prey and predator populations determine that this system moves to the steady state through a series of damped oscillations, the predator curve lagging behind the prey curve (Figure 1.2b). Secondly, system state may move away from a point in the phase space, in which instance the system is said to be unstable (Figure 1.2c). In the predation example, this would result from an over-efficient predator which is capable of exploiting its prey at very low prey densities. This system becomes unstable through a series of oscillations of increasing magnitude

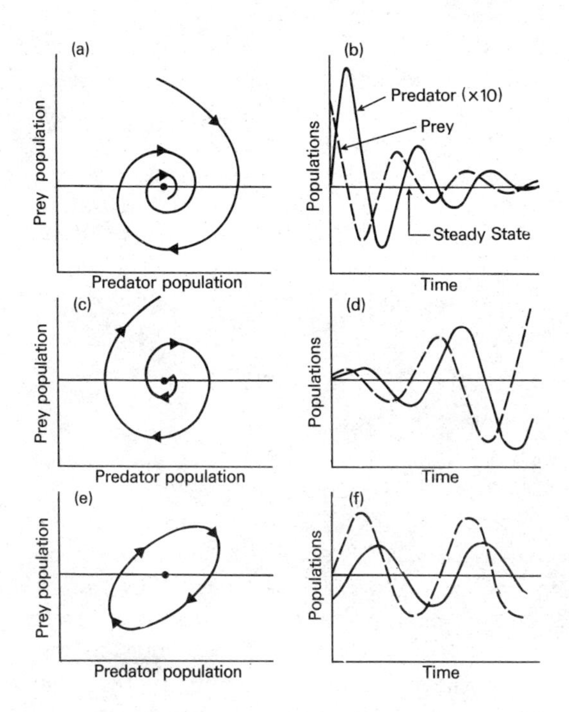

FIG. 1.2. Types of system stability shown by a predator–prey system. The figures on the left show the system dynamics within a phase plane; the figures on the right show system changes through time: (a) and (b) show stable dynamics; (c) and (d) show unstable dynamics; (e) and (f) show cyclically stable dynamics.

and eventually the prey population will become extinct and, unless alternative prey should be available, the predator population will soon follow suit (Figure 1.2d). Finally, system state may circle a point in the phase space, in which case the system will exhibit periodic or cyclical fluctuations about a steady state (Figure 1.2e). In the predation example this situation represents a stable state in which both prey and predator populations show regular fluctuations through time, the changes in the predator population lagging behind the changes in the prey population (Figure 1.2f); this cyclical change in population numbers is an immanent system property and has nothing to do with seasonal variations in numbers.

A phase plane may be visualized as a land surface. A stable point is represented in the landscape by a pit or depression into which a ball placed nearby on the surface will roll (Figure 1.3a). A stable point may also lie on a watershed where a ball will remain so long as it is not subjected to any displacements; should the ball be displaced, it will move away from the stable point (Figure 1.3b). The ball in the pit is said to be

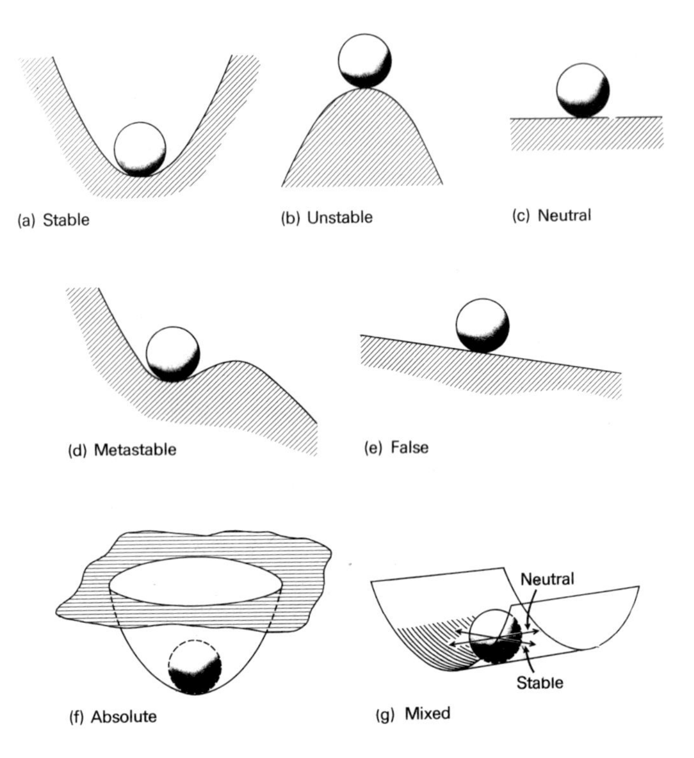

FIG. 1.3. Types of system stability: mechanical analogies. Reproduced with permission, from D. C. Spanner (1964), Figure 8.1 (p. 99), *Introduction to thermodynamics*, Copyright by Academic Press Inc. (London) Ltd.

in stable equilibrium whereas the ball on the watershed is said to be in unstable equilibrium. At a metastable point (Figure 1.3d), the ball will stay in its pit for a limited range of displacements. Should a threshold level of displacement be reached, the ball will move off to another part of the landscape. The landscape of the phase plane, its equivalent in our n-dimensional phase space, may be very contorted, containing equilibrial nodes of all varieties (stable, unstable, metastable, and so on); this phase-plane complexity has important implications to system behaviour. Gray (1976), for instance, has shown that in some marine systems a global stability is often assumed for the entire phase space, with the one stable point representing a climax community (Figure 1.4a), and no matter how much the system is disturbed it will always return to a stable equilibrium condition dominated by the same species. More realistic, according to Gray (1976), is the idea of neighbourhood stability in which different levels of stability are possible in the system which depend upon the level of disturbance. Thus in Figure 1.4b, a certain species is locally

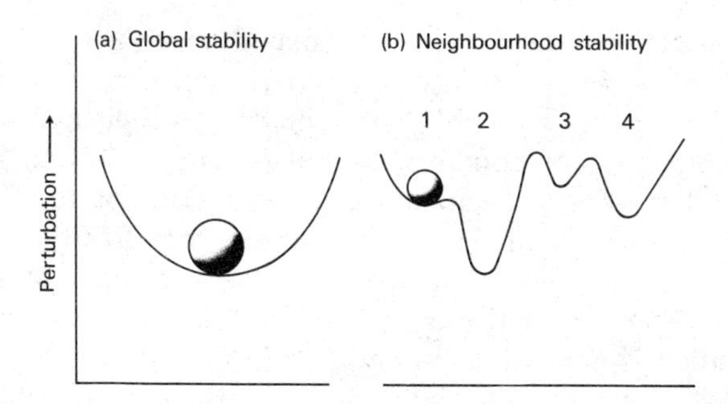

FIG. 1.4. Neighbourhood stability. Reprinted with permission, from J. Gray (1976). This first appeared in *New Scientist*, London, the weekly review of science and technology.

stable in point 1, but a slight disturbance may send the system to the stable state at point 2 which is dominated by a different species. A larger disturbance would then be required to shift the system to the stable state at point 3, and so forth. This notion of local stability would seem appropriate for the stability characteristics of most systems of geographical interest which, owing to constantly changing inputs, tend to wander through successive stable points in their phase spaces. Until recently, phase-space studies assumed a continuous 'surface'. Recently, Thom (1972) has developed a topology of catastrophe theory to permit sudden discontinuities in space and time including shock waves, embryonic development, and many human relations. In essence, discontinuities allow the system to flip from one domain of operation to another. Mathematical discussion of catastrophe theory is given in Bennett and Chorley (1978) while Kirkby (1978) furnishes some interesting examples of phase-plane behaviour from physical geography.

The terms 'stable', 'equilibrium', and 'steady state' have turned up in discussion; unfortunately, the usage of these words is not unequivocal and needs a little explaining. The meaning of the term 'equilibrium' in a mechanical system is easy to grasp: it indicates a state of balance, of rest, of absence of change between different components or elements operating in a given situation. If a boulder should fall from a crag, it will roll down a scree slope and eventually come to rest: it will then be in equilibrium. Applied to systems other than mechanical ones, equilibrium must satisfy two conditions. Firstly, the state variables should not vary with time. These variables are macroscopic; a system in equilibrium would be permitted to possess movement at the microsopic level where Brownian movement—the random motion of molecules—prevails. Secondly, the system must also be at rest in the sense that no flow of matter or energy is maintaining the constancy of the state variables; this means that the components of the system are isolated from the environment—no cross-boundary exchanges are allowed. Thus, an organism on reaching maturity will maintain a roughly constant body weight and hence fulfils the first requirement of equilibrium; but since it needs to import and export quantities of matter to maintain this state, it fails the second requirement of equilibrium. The state of any system, like the organism, in which a balance exists but changes do occur, is referred to as a stationary state or steady state. Rayner (1972) explained that, because in all natural systems isolation is impossible and equilibrium is a theoretical condition, most uses of the term 'equilibrium', particularly in geomorphology, refer to a steady state; and that, fortunately, so long as workers are aware of this loose meaning of equilibrium, it should not lead to misunderstanding. Unfortunately, the problem is compounded by the notion of dynamic equilibrium. As has been stated, for a system to be in equilibrium there should be no macroscopic processes taking place, though microscopic processes are permitted. Thus, in thermodynamics, where molecules move chaotically, even in equilibrium, the equilibrium is clearly a dynamic affair: unlike the components of a mechanical system which are at rest in an equilibrial state, the molecules in a thermodynamic system in equilibrium move—but their movements balance out. In geography, alas, the term 'dynamic equilibrium', instead of being applied in this restricted sense, is taken either as synonymous with 'steady state' or with some kind of false equilibrium in which the system appears to be in equilibrium (*qua* steady state) but in reality is changing very, very slowly with time. In brief: a steady state is a dynamic condition but at the macroscopic level; equilibrium, in the thermodynamic sense is also dynamic but at the microscopic level only; dynamic equilibrium is often applied to macroscopic changes in systems of geographical interest. Readers who wish to arm themselves with further jargon of system dynamics should consult the relevant chapter in Chorley and Kennedy (1971).

1.3. The classification of systems

Systems may be classified in many ways. For instance, the division of systems into isolated, closed, and open is a functional classification. The classification of system types is one of the most confusing parts of the systems literature. Owing partly to the latitude of interpretation as to what a system is, by the array of disciplines which address themselves to the definition of systems, there is a terminological labyrinth on the subject. Trying to construct an over-all classification of system types is a Herculean task. An attempt to do so has been made by Ackoff (1971) but his scheme contains much jargon and categories such as goal-seeking systems and ideal-seeking systems sound like technical descriptions of footballers and perfectionists respectively. Rather than getting bogged down in this largely non-geographical literature, much of which seems ripe for a good debunking, a home-grown classification due to Chorley and Kennedy (1971) will be briefly discussed. Chorley and Kennedy (1971, p. 3) gave a structural classification based on the degree of internal system complexity; they suggested that the complexity of natural systems increases roughly as follows. The simplest structures are found in *morphological systems* which consist of a web of relations between system parts, the strength of the relations being given as some measure of statistical association. Figure 1.5 shows a simple interfluve slope system on sandy limestone in eastern France represented as a morphological system. The relation between system components are measured by a correlation coefficient and have both strength and direction (positive or negative values). At the second level of internal system complexity are found *cascading systems* in which system parts are linked by a flow of mass or energy or both. An example is an erosional drainage basin system in which debris and water are shunted from the valley-side slope sub-system to the stream channel sub-system. The third class of systems are called *process–response systems* which are characterized by at least one link between a morphological variable and a flow in a cascading system. Thus a simple process–response system would be a splash pit produced by a falling raindrop in which a morphological variable, say splash-pit depth, is related to an energy variable, say the kinetic energy of the falling raindrop. Process-response systems stress the relations between system *form* and system *process*. At the fourth level of internal-system complexity are *control systems;* these are process–response systems in which the key components are controlled by man. Chorley and Kennedy

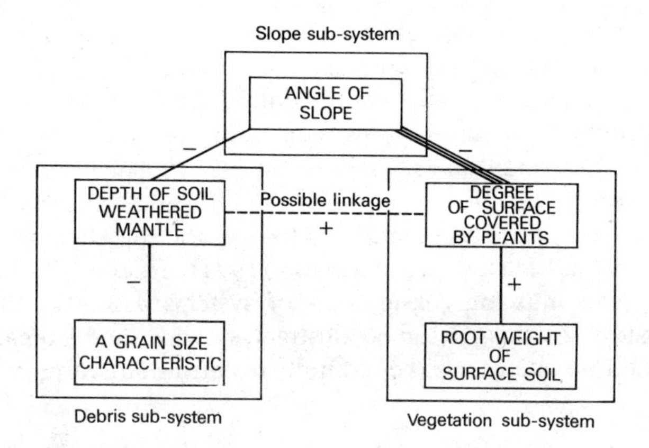

FIG. 1.5. An interfluve slope represented as a morphological system. Reprinted with permission, from R. J. Chorley and B. A. Kennedy (1971).

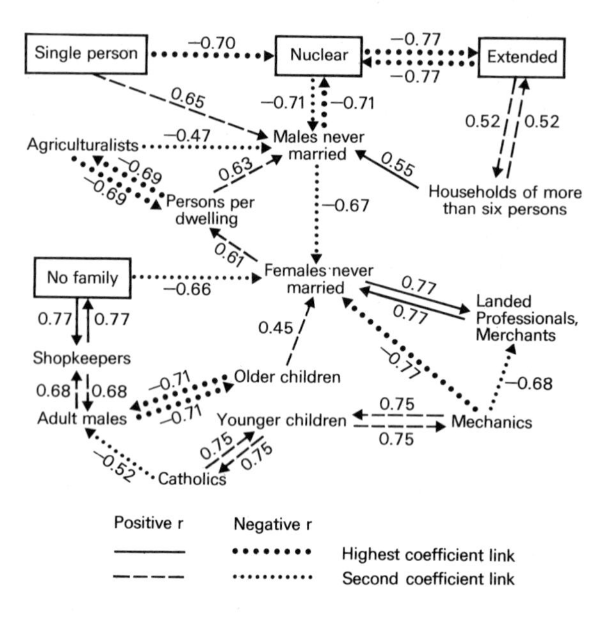

Fig. 1.6 The structure of eighteenth-century New Zealand households represented as a morphological system. Reprinted with permission, from P. H. Curson (1976), 'Household structure in nineteenth-century Auckland', *New Zealand Geographer*, **32,** 177–93.

then list several other classes starting with self-maintaining systems,[1] as represented by the simplest forms of life, and running through plants, animals, ecosystems (plants, animals, and their inanimate environment), man, social systems, and ending with human ecosystems (socio-ecological complexes). In fact all these biological, ecological, and social systems could be viewed and studied as morphological, cascading, process–response, or control systems. For example, Figure 1.6 shows the social structure of households in nineteenth-century Auckland represented as a morphological system. And an ecosystem could be viewed as a cascading system in which energy and mass pass through successive trophic levels (sub-system units).

It might be useful at this juncture to look at the so-called systems map of the universe as charted by Checkland (1971) (Figure 1.7). Checkland divided his map into four basic areas of systems: natural systems, designed physical systems, designed abstract systems, and human-activity systems. Natural systems are the naturally occurring structures which man

[1] Since many non-living systems can maintain themselves—a thunderstorm will serve as an example, the expression 'self-replicating systems', by limiting examples to the biological realm, would seem more apposite here.

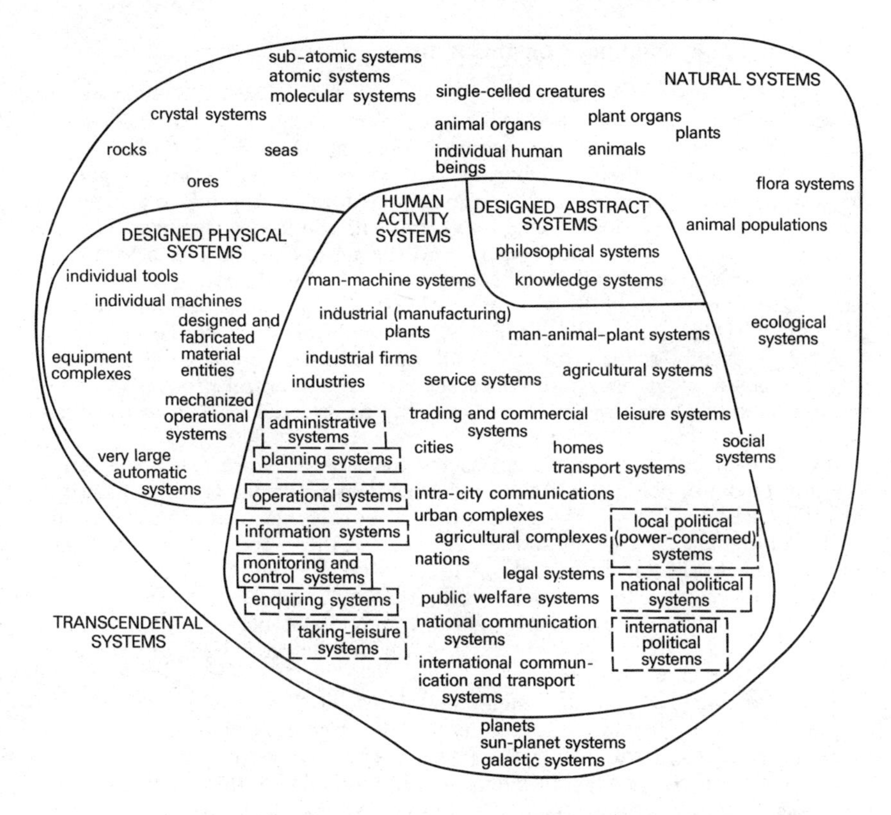

FIG. 1.7. Checkland's (1972) systems map of the universe. Reprinted with permission, from P. B. Checkland (1972).

has identified within the universe. Designed physical systems do not exist in the natural world but are needed in a human-activity system. Designed abstract systems are systems of thought or philosophies. Human-activity systems result from man exercising his abilities as tool-maker, myth-maker, and doer to change his environment and his circumstances in his environment. Notice that on the map, social and cultural systems bestride the boundary of natural systems and human activity systems; this is because social and cultural systems provide a framework within which nearly all human activity systems exist but, at the same time, they are natural systems in so far as man can be regarded as a social animal population.

1.4. Systems of interest to geographers or geographical systems?

Let us now turn to the question of what systems are of interest to geographers. Wilson and Kirkby (1975, p. 4) have made a lucid and unobjectionable statement on this matter:

> The systems of interest to geographers can be divided almost unambiguously into those of human geography and those of physical geography, but further subdivision involves alternative overlapping concepts. Human geography is concerned with people, their activities, and their spatial distribution (population geography, social geography), with organizations formed to produce a variety of goods and services (economic geography), and with the use of resources (resource geography, economic geography again, and specialized subjects such as agricultural geography). Spatial analysis is related to various forms of regions, and particularly important units are cities, or urban regions (urban and regional geography). There is also an older (but still important) use of the concept of regional geography as the synthesis of all aspects of the geographical study of a region. Human geographers are concerned with geographical processes in time, and in relation to long periods, such work may be referred to as historical geography.
>
> Physical geographers study landforms and the associated processes of sediment transport (geomorphology); and the spatial relationship of plants and animals and their interaction with soil (biogeography). They are also concerned with the pattern of climate on a world and local scale, and the processes which influence climate on a world and local scale, and the processes which influence climate, especially in relation to the soil and to the hydrological cycle (climatology, micrometeorology). Water is an important component of most physiographic systems and the patterns and processes governing its movement are therefore also of critical interest (hydrology).
>
> There are potentially strong interactions between these two major fields of geography. Broadly speaking, we can say that the structure of human activity in a region is in part determined by the physical environment in that region. The economic structure is partly determined by the distribution of natural resources. Water resource management is a function of climate and the structure of the local river network. The interdependence is probably strongest in less developed societies. In countries with advanced technologies, man's control over major parts of his environment in cities to a major extent decouples (at least for analytical purposes) the human and physical systems. It is now increasingly argued, however, that in advanced technological societies, there are strong interconnections, the study and planning of which have been neglected, and that this neglect will lead to serious repercussions—at the extreme, the so-called eco-catastrophies. Geographers, and others therefore, are increasingly involved in the study of this kind of ecology.

Wilson and Kirkby use the phrase 'systems of interest to geographers', and thus far in this book we have used a phrase of that ilk; this avoids using the contentious phrase 'geographical systems'. Do geographical systems exist or are they chimeras? There is perhaps a danger here of making much of a problem which does not exist. It could be argued that if there is a set of systems, like that outlined by Wilson and Kirkby, in which geographers are interested, and if the definition of these systems is beyond dispute, then all is well and each geographer may happily plough his geographical furrow in a field of science. It would follow from this argument that, as we have often been reminded, geography is the doings of geographers. Some would declare that these doings were concerned with spatial relationships and in this wise different from the doings in cognate disciplines. On the other hand, there may be entities which by their relations form distinctly geographical systems. This question has been touched on by Chisholm (1975, p. 35) who observed that there is in geography, and in any other discipline for that matter, a spectrum of systems, ranging from those that are readily identifiable, such as a river drainage basin, to those which are susceptible of much controversy, such as a city region. Fuzzy notions of geographical systems, and even the overt denial of their existence, like the difficulties in seeing biological units above the level of individual organisms, may stem from the geographer's understandably anthropocentric view of the world. Laszlo (1972, p. 167) made a useful distinction between our perception of things on either side of what he called, meaning ourselves as human subjects, the conceptual rubicon: we tend to see smaller systems than ourselves—other organisms, cells, atoms, and so on—as distinct objects, whereas in those larger systems of which we are ourselves but a part—social and ecological systems—we tend to see relationships between things and not a unitary system. Certainly, one of the reasons geographers are unhappy with the concept of, say, a regional system, is because, unlike other organisms or crystals, these systems have no clearly demarcated boundaries, a point to which we shall return in chapter four.

An attempt to define geographical systems was made by Huggett (1976). Laszlo (1972) had argued that the multifarious systems of which the universe is composed—atoms, planets, galaxies, cells, organisms, societies, and so forth—have a basic twofold, hierarchical evolution. On the one hand evolve the entities of astronomy: galaxy clusters, galaxies, star clusters, stars, and planets and their subsidiary bodies. On the other hand evolve the entities of physics, chemistry, biology, ecology, sociology, and international organizations: atoms, molecules, molecular compounds, crystals, cells, multi-cellular organisms, and communities of organisms. The basic building blocks of both hierarchies are atoms, though in turn atoms are composed of elementary particles, which themselves may be constituted of quarks, and all of which presumably exist in a sort of space-time manifold. To simplify the terminology Laszlo (1972, p. 28) had coined the term 'microhierarchy' for the terrestrial-type atoms-to-ecologies hierarchy (to be more precise, this should be called the atoms-to-societies hierarchy) and, for the astronomical hierarchy he had coined the term 'macrohierarchy'. Huggett (1976) revised and extended Laszlo's bipartite, linear scheme to include another microhierarchy which provides an evolutionary link between atoms and planets in the macrohierarchy, namely, the hierarchy of planetary and geological systems, or 'atoms-to-planets hierarchy', as it might be nicknamed. The important distinction between inorganic and organic terrestrial hierarchies had been made by Gerard (1969) who defined two classes of systems, or 'orgs' as he had styled them: non-living orgs which form an evolutionary line from atoms, through molecules and larger units, to geographical, geological, and astronomical units; and living animorgs which form an evolutionary line from unicellular organisms, through multi-cellular organisms, to epi-organisms such as ant-hills and villages. It was noted by Huggett (1976) that the systems of the atoms-to-planet and atoms-to-societies hierarchies commingle to produce a third microhierarchy: the hierarchy of environmental systems. This third hierarchy might better have been named the 'atoms-to-ecologies hierarchy'.

Huggett (1976) construed geography as that science which deals with systems at the uppermost levels of the atoms-to-ecologies hierarchy, though some geographical practitioners delve into systems at a lower level. His argument was that geographical systems are formed by the interaction of social, ecological, geological units, and so on, and this interaction is governed by geographical laws. Partial support for this view can be mustered from the writings of other geographers. Anuchin (1973) for instance, saw the geographical sphere as a complex of systems—lithosphere, hydrosphere, atmosphere, biosphere, and sociosphere—which develop and endure through natural interaction of physical chemical, biological, geological, and social phenomena. And Chorley and Kennedy (1971, pp. 341–3) argued that socio-economic systems, which are dominantly non-spatial, interact with physical process–response systems, which tend to be spatial, to form at intersections between the two the stuff of geography. Where interest is centred upon process-response systems alone, the subject matter is that of physical geography; where interest is focused on spatial aspects of socio-economic systems, the subject matter is that of human geography. They contended that, in many cases, physical process–response systems make a small but necessary contribution to the creation of a truly geographical system. For without the inherently spatial physical component, the whole system would consist of social and economic variables whose specialized locational and geographical components would receive scant attention other than regarding regions as little boxes subject to inputs and outputs. By being explicitly geographical units with a definite spatial magnitude and location, physical systems, when linked with socio-economic systems, impart to the latter a distinctly regional and locational flavour which is palpably geographical.

2 What is Systems Analysis?

> Now, whoever will please to take this scheme, and
> either reduce or adapt it to an intellectual state
> or commonwealth of learning, will soon discover the
> first ground of disagreement between the two great
> parties at this time in arms, and may form just
> conclusions on the merits of either cause.
>
> SWIFT, *The Battle of the Books*

SYSTEMS analysis in geography is in a nebulous state, as the following quotations suggest:

> Nevertheless the fact remains that, despite its present limitations, both the frame of mind induced and the research prospects held out by systems analysis require that this approach must be seriously explored as the major methodological effort of geography during the next few years. (R. J. Chorley (1973, p. 166))

> Altogether it seems likely that by the end of the present decade it will be generally accepted by geographers that while systems, like regions, provide a useful framework within which to work, they are all too frequently intangible things that with maddening regularity retreat from the researcher—just as the bag of gold at the rainbow's end eludes the seeker after riches. (M. D. I. Chisholm (1975), p. 36))

This may be in part owing to the ambiguous and sometimes emotive usage of the phrase systems analysis. For, it taken literally, and despite the apparent triteness of the statement, systems analysis is the analysis of systems, a system being broadly defined as a set of interrelated parts. And systems analysis in geography is the analysis of systems of interest to geographers like drainage basins or city regions. All geographers analyse some system or other, so, without knowing it, all geographers have engaged in some form of systems analysis. Or have they? Perhaps the essence of systems analysis is not captured by its literal definition. Perhaps a more restricted definition might exclude from systems-analytical activity certain modes of study. We might resolve this poser by asking what, if anything, is the difference between traditional scientific enquiry, which engages the analysis of a system, and systems analysis, which seemingly sets about the same task? If there is no difference between the two, and on the face of it there is not because all scientists study some system or another, we may reasonably ask why use the term 'systems analysis'?

A careful comparison between systems analysis and the procedures of traditional science would suggest that there is no difference between the two. For, as practised, systems analysis usually involves the formulation of a model of a complex system. This requires both the specification of real-world variables and statement of hypothetical relations among these variables. As Ashby (1966) pointed out, any real system is characterized by an infinity of variables and different observers with different, or perhaps even the same, aims may make an infinity of selections. Or, as Chadwick (1971) put it, the detection of a system is up to the researcher and is thus subjective—a system, like beauty, lies in the eye of the beholder. A model of a system is thus a hypothesis, predictions from which can be tested against real-world measurements.

Depending on the agreement between observed and predicted results, the model will be accepted, rejected, or modified in some way, new predictions made, and retested. Systems analysis is thus an iterative process akin to the traditional method of empirical science. But notice we stated that the model is of a complex system. Does then the distinction between systems analysis and normal science lie in the complexity of the system studied? To answer this question we must discuss how complex systems are distinguished from simple systems.

2.1. The simple and the complex

In a simple mechanical system, as studied by Newton, the motion of each system component—let us take the example of a few molecules of hydrogen moving in a confined space—can be described by deterministic equations of motion, that is, equations which define the exact movement of the particles with respect to the forces acting on them. Taken together, the equations for individual hydrogen molecules form a set of simultaneous equations which enable us to predict, for a given initial distribution of molecules, how the system will change with time, to state precisely the future locations of all the molecules. Now physicists found that some mechanical systems are too complex to tackle by the orthodox Newtonian approach of finding solutions to the deterministic equations, too complex that is in terms of the sheer number of components in the system: a handlable quantity of hydrogen gas contains roughly 10^{23} molecules, and the solution of 10^{23} simultaneous equations is clearly utterly impractical. What physicists did, therefore, was to explain known laws relating *macroscopic variables* (these variables include temperature and pressure and are suitable for describing properties of large quantities of gas) in terms of the component *microscopic variables* obeying the known laws of the mechanics of motion. The result is that physicists have statements about the statistical properties of large populations of particles, such as the Maxwell–Boltzmann Law; and statements about macroscopic observables, for instance pressure is proportional to the product of the average number of particles in the gas, their mass, and their average velocity.

These achievements of what has become known as statistical mechanics depend on finding some conserved, macroscopic quantity which remains constant throughout the microscropic motion of particles.

Does this kind of thinking apply to systems of interest to geographers? Certainly, geographically interesting systems are invariably complex in the sense that, if analysis is carried out down to microscopic levels, the system components are countless and the interaction between them enormously rich. The geomorphologist interested in large-scale landscape systems cannot hope to consider the behaviour and interaction of each and every slope and river in the landscape. Nor can the human geographer interested in population movements within a large, or even medium-sized region, hope to model the movement of every individual within the population. So can geographers get round this problem by finding, as in statistical mechanics, a conserved, macroscopic quantity which remains constant throughout the microscopic behaviour of system components? Or, what is the same thing, can they find macroscopic variables which are opposite to the description of whole landscapes or populations or the state of any other system of geographical interest? To answer this question we must interpret the meaning of 'microscopic' and 'macroscopic' in a geographical setting. Useful in this regard, is the idea of levels of resolution in systems. All systems, as we have seen, seem to have a hierarchical structure in so far as they can be broken down into sub-systems, each of which in turn may be further subdivided into smaller sub-systems. Thus, for instance, a social system can be viewed or resolved at a number of levels ranging from, at the lowermost level, individuals, through intermediary units such as tribes, to nations and interational organizations. It can be seen that systems at one level of resolution are the components of systems at a higher level of resolution. Thus is demography, individuals might be thought of as the microscopic components of human populations and populations densities of individuals the macroscopic variables. Thus by discriminating system parts of sub-systems at different levels of resolution, a complex system is simplified in a logical and realistic way which avoids the bewildering bulk of information at microscopic levels. The process involves viewing the system through Odum's (1971) macroscope which, instead of focusing on detail as a microscope would, has its maximum resolving power on the general structure of the whole system and can also zoom in and scan intermediary system parts (sub-systems) at intermediary levels of resolution.

2.2. Ways of analysing complexity

The problem of applying the statistical mechanics method to geography now seems to have a straightforward solution: define macroscopic variables for systems of interest to geographers, establish laws between these variables, and we are just about home and dry. Alas, this is not so easily done, for there are few, if any, generally accepted, macroscopic variables suitable for describing geographically interesting systems. Different schools of thought exist in geography which, in essaying the task of describing and analysing complex systems, have come up with their own particular methods of assault (Table 2.1). To dub these schools 'the theoretical bunch' and 'the empirical mob' would do injustice to both for the protagonists of each form a very mixed bag and in many ways their work is complementary; but, polarizing the two schools, that is the difference between them.

The theorists, and there are not many of them, draw their theory from many sources: from statistical mechanics, from Newtonian mechanics, from sociology and economics; there are even a few theories, such as those of W. M. Davis, which are indigenous to geography. It is useful and logical to divide theoretical analysis into two types which differ in the level of system resolution for which a theoretical model is set up: models based on *micro-scale theory* and models based on *macro-scale theory*. In models based on micro-scale theory, only microscopic system parts are of concern and there is a one-to-one or isomorphic correspondence between basic parts of the system and sub-system models. For instance, in spite of the likely unwieldiness, it would be possible to establish in theory the energy flow through, say, a woodland ecosystem which incorporated every species present. In models based on macro-scale theory, in which low-resolution system parts are grouped into sub-system units and theory concerns the relations between grouped components, there is a many-to-one or homomorphic correspondence between basic parts of a system and sub-system units. In the case of the woodland ecosystem, species might be grouped into units on the basis of feeding habits—herbivores, carnivores, and so forth, and a more manageable model of energy could be developed from a theoretical consideration of flow and storage in the sub-system units. It is interesting to note that, despite these differences, the behaviour of both microscopic and macroscopic systems can be mathematically given the same general formulation and both systems share, in part, the same general properties.

Table 2.1
Ways of analysing systems

Level of resolution	Type of Analysis *Theoretical*	*Empirical*
High (macroscopic)	Macro-scale theory	Experimental component method
Low (microscopic)	Micro-scale theory	Traditional empirical method

The other school of thought differs in the way in which relations are established between system parts. In the absence of microscopic and macroscopic laws, one can observe the nature of the relations between system parts and this is what this second school does. Within this school there is, however, an important split which is largely attributable to the level of resolution of parts for which empirical relations are sought. On the one hand, one may seek relations at lower levels of resolution; this has been the regular practice, especially in physical geography, where, notwithstanding some important theoretical work, systems of interest have been minutely dissected and studied by field observation or experiment or, less commonly, by laboratory experiment. This traditional empirical method of analysis is particularly efficacious if just few low-resolution parts are studied over a short period of time, in which case the observed system changes can often be described by relatively simple equations. Thus, in hydrology, some workers have carried out detailed research into individual components of the basin water cycle. For instance, Penman has studied evaporation and developed an empirical equation for its prediction which uses meteorological variables; and Horton has studied infiltration and set up an empirical mathematical model which, given conditions such as soil wetness, vegetation cover, and length of storms, can be used to predict infiltration rates. A major limitation of this method is that the equations relate to but a small part of the whole system, about the behaviour of which whole the equations say little or nothing. It has been said that the ultimate aim of this, the physical-science approach to analysis, is to give a rational synthesis of a system through an understanding of all its physio-chemical mechanisms and interactions. However, all attempts to piece together small-scale relations to produce a set of relations which explain the behaviour of the whole system are doomed to failure because they look over the relations which exist between system units at a higher level of resolution; the inevitable outcome of such an attempt is a hotch-potch of small-scale predictions, which may in themselves be of interest but have only an indirect bearing on what the system as a whole does.

On the other hand, relations may be sought between sub-systems at higher levels of resolution. Emphasis here is thus on obtaining workable relations between high-resolution sub-systems which can be used to predict the behaviour of the system as a whole. In hydrology this approach is exemplified by the work of Crawford and Linsley (1966) who represent the whole of the land phase of the water cycle by a series of storage components, through which incoming water is routed. This model has been very successfully used to predict the behaviour of water at the level of the whole basin system. Since the relations between system components—the storage components in the hydrological example—are usually established by experimental work, this type of empirical analysis might be termed the *experimental component method*, following the usage of this phrase in ecology.

Of course, the differences between these four kinds of analysis are seldom clear cut. In practice, these methods of analysing systems will partly complement one another, notably in the horizontal sense between theory and observation. As a procedure, systems analysis should have both theoretical and empirical facets: were it not for field measurements and testing, theoretical analysis would become increasingly abstract; were it not for theoretical models, it would be difficult to see any general application of particular observations.

Returning now to the discussion about what constitutes systems analysis, we may ask if all the analytical categories shown in Table 2.1 are part of systems analysis in the more restricted use of the term, that is, the analysis of complex systems. From our consideration of complex systems the answer is no. There are technical differences between microscopic analysis and macroscopic analysis and to many workers systems analysis is macroscopic analysis. Nevertheless, as we have noted, the distinction between on the one hand what we have styled micro-scale theory and the traditional empirical method, and on the other hand their macroscopic counterparts, macro-scale theory and experimental component method, is not always as sharp as the table suggests. We shall therefore in this book include mention of the microscopic models of enquiry wherever they are pertinent to the understanding of problems at the macroscopic level.

2.3. The strategy of systems analysis

It matters not whether one wishes to pursue a theoretical or an experimental mode of analysis; in both cases the strategy of systems analysis will engage the same four phases, viz. the lexical phase, the parsing phase, the modelling phase, and the analysis phase. Most books concerned with systems and their analysis would at this juncture depict some more detailed variant of these phases in the form of a flow diagram (see, for example, Trudgill, 1977, p. 8); this gives the impression of a mechanical procedure lacking in flexibility. We shall, therefore, simply leave the procedure in the loose frame of four broad phases and consider each in turn.

The lexical phase is analogous to the establishment of a vocabulary for a language inasmuch as it includes the recognition of basic system components; it entails three steps. The first step is the defining of hypotheses to be tested or questions to be answered about a system of interest. The second step is the placing of bounds on the system of interest; this is known as system closure and requires the separation of the system and its environment. The third step is the choosing of state variables which will define the state of system components. The parsing phase is analogous to the establishment in a language of grammatical rules which govern the relations of words, one to another; it involves the defining of relations between systems components. These relations may take the form of theoretical equations, empirical equations, or simply correlation coefficients. The modelling phase has two steps; firstly, mechanisms whereby the relations act upon the state variables to effect changes in system state are elucidated; in other words, the system components and measures or statements of their interaction are put together to form a model of some sort. Secondly, the model is operationalized or calibrated by giving actual values to parameters and constants. In the analysis phase the system model is solved, at least sometimes it is, to produce assessable results. If the results are not good enough, which usually means they match poorly with observed changes in system state, the procedure starts again, suitable modifications to the model being made.

This procedure should come to life when applied to an actual case. Take the study made by Hett and O'Neill (1974) of the Aleut ecosystem. The Aleuts, before contact with civilization, were a marine hunting and gathering people, very dependent on the ocean for food and clothing, while the terrestrial system provided food supplements. Questions which might be answered by systems analysis are: (1) How dependent were Aleuts on marine and on terrestrial components of the Aleutian ecosystem? (2) Did the Aleuts' using the marine environment stem from inherent properties of the system which made it advantageous for them to do so? (3) How important were the Aleuts in influencing the stability of the entire Aleutian ecosystem? The boundaries of the system are easily set: the ecosystem is a linear island chain, some 2016 km long and 65.5 km wide, running from Port Moller on the Alaskan Peninsular in the east to Attu island in the west. The components of the system are shown in Figure 2.1 and are defined by the amounts of carbon in store. (Many studies of ecosystems use energy stored but carbon has the advantage of being an essential component in living organisms in the system, through which it flows steadily.) All this comprises the lexical phase. The parsing phase seeks relations between state variables; in the Aleutian case we are looking at carbon moving through the system. The relations are thus actual flows of carbon (see Figure 2.1). In the modelling phase the state variables and their interaction must be joined to form a model.

In the Aleutian case, a mathematical model may be formulated as a set of equations describing the change in storage of carbon in each state variable over a small time interval as a function of carbon inputs and carbon outputs. To solve the equations so that the value of each state variable at any point in time can be found, it is necessary to know the

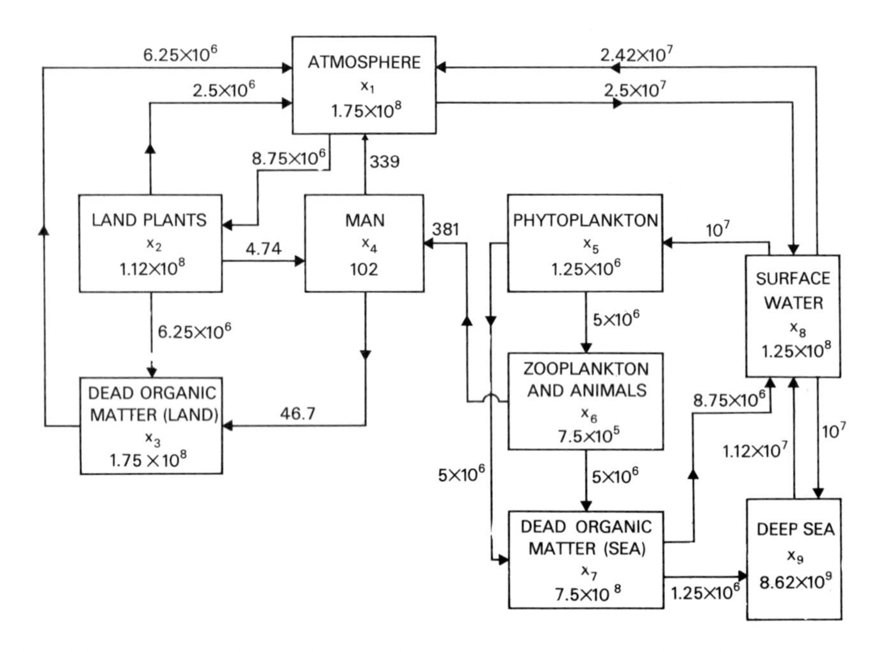

FIG. 2.1. Carbon model for the Aleutian ecosystem. The values in each compartment represent metric tons (10^3) of carbon at steady state. Numbers on the arrows represent the flux of carbon (metric tons per year) between compartments. Reprinted with permission, from J. M. Hett and R. V. O'Neill (1974), 'Systems analysis of the Aleut ecosystem', *Arctic Anthropology*, **xi. 1** (© 1974 by the Board of Regents of the University of Wisconsin System), 31–40.

initial contents of carbon in the system components and the rate constants determining flow rates of carbon between components; the measuring of these quantities enables the model to be operationalized (see Hett and O'Neill (1974) and chapter 6 below for details). Having set up a model of the Aleut ecosystem, a variety of analytical techniques may be used to address the questions posed at the start of the lexical phase. Without going into technical detail, analysis showed that the model faithfully represented the general nature of the Aleut ecosystem; that the Aleuts are dependent on a marine food supply but far more so than was originally thought; that the system is free from large-scale disturbances by virtue of its internal buffering mechanisms rather than through a lack of external disturbances such as fire and drought; and that the Aleut population has little effect on the stability of the total system. As Hett and O'Neill (1974) stressed, this is a preliminary analysis of the Aleut ecosystem which should be followed up with further cycles of systems analysis.

3 The Geographical Case For Systems Analysis

> We system-makers can sustain
> The thesis that you grant was plain
>
> PRIOR, *Alma*

3.1. The promise of systems analysis in geography

THE geographical case for systems analysis is a compelling one; there are at least five good reasons why geographers might usefully use systems-analytical methods. Firstly, the analysis of complex systems, generally avoided by scientists until recently, seems beyond the capabilities of traditional scientific methods. New techniques have been developed outside geography to tackle complex situations and these techniques are the tools of systems analysis in its broadest sense (Waddington, 1977). Since systems of interest to geographers are usually complex ones, systems analysis should appeal to the geographer. Indeed Harvey (1969, p. 487) wondered how geographers, given the complexity and richness of interaction with which they deal, could avoid using techniques and terminology specifically devised to attack complexity. Secondly, the notion of levels of resolution and the explicit statement of hierarchial structure incorporated in systems analysis should provide an appropriate methodology for a fresh attempt at assailing the so-called scale problem in geography. The scale problem is that an explanation of system dynamics at one level of resolution will seldom apply to system dynamics at a higher level or lower level: a knowledge of why people move within a small area during a relatively short time does not enable longer-term population movements over larger distances to be explained. Most systems consist of sub-system units arranged in a hierarchy. Although these sub-systems operate at various space and time scales, they do so conjointly. The systems approach offers a framework in which to elucidate the mechanisms whereby the dynamics of various components in a hierarchy are linked; for instance, the mechanisms which link sub-systems at different levels in the administrative decision-making hierarchy in England and Wales—parish and borough councils, county councils, and central government. Thirdly, the openness of system units within a hierarchy, such as an urban hierarchy—cities, towns, villages, hamlets—is embodied in a systems approach. Any system is seen to have a dynamical relation with its environment and with other system units from the same hierarchy: it receives energy, (solar energy, fossil fuel, and so on), matter (goods, minerals, and the like), and information (personal contract, newspapers, and so forth), transforms and stores them, and exports them. This framework seems apposite to all systems which interest geographers.

Fourthly, the possibility of establishing macro-variables and macro-laws, analogous to gas laws is statistical mechanics, has a certain appeal and might well be one path to geographical theory. Some progress has been made in this direction in both physical geography, where the height of the land surface has been proposed as a macro-variable of the landscape (Scheidegger, 1970), and in human geography, where income potential has been put forward as a macro-variable of human populations (Warntz, 1973). Furthermore, in both these cases tentative macro-laws have been suggested and we shall look into these later in the book.

Fifthly, the systems approach imparts a terminological coherence to the whole field of geography, the jargon being translated into the lingua franca of much of natural, physical, and engineering science with which subjects communication is thus improved (Cooke, 1971).

3.2. The present status of systems analysis in geography

Despite its promise in geography, systems analysis has not yet achieved anything approaching the operational status its intrinsic merit would seem to warrant. Held by the few geographers as a recipe for great theoretical and methodological advances, balked at by the many, its direct value to geographical practitioners has yet to be convincingly demonstrated. Harvey's (1969) remark, that, for the most part, systems analysis in geography has not advanced much beyond the stage where we are exhorted to think in terms of systems, still applies today: progress has been exceedingly slow. It is fairly easy to pinpoint the factors which explain the geographers' reticence to adopt system analysis. The main factor seems to be the sheer complexity and structural richness of systems of interest to geographers, which are replete with non-linear relations, complicated by lags, variable thresholds, and a propensity to lurch from one state of disequilibrium to another (Chorley, 1973). This factor deters geographers for two reasons: firstly, the mathematical formulation of systems studied by geographers is somewhat tricky and would certainly be shunned by the mathematically shy geographer; secondly, the modelling of complex systems, especially those which come within the ambits of human geography, requires an evaluative judgement in closure and variable selection which, if it is to be done with confidence, needs much experience (Harvey, 1969); and also requires that component parts of the systems should be studied in a manner unfamiliar to most geographers, that is, a functional manner (Langton, 1972). Another contributory factor, noted by Langton (1972), is the problem that all social systems contain a duality between things and images of things, between the real and the perceived. However, mathematical descriptions of human personality and interaction are being evolved (Clymer, 1972) and perhaps we should regard this as a path to explore rather than a dead end. A third factor, again mentioned by Langton (1972), is the problem of system closure, that is, deciding on what is system and what is environment in the intricate and highly interconnected milieu that geographers study. Finally, the systems approach itself is not without flaws: Hoos (1972) saw both strength and weakness engendered by the multifarious forms and manifestations of the systems approach, the very variety of which perhaps stems from the latitude of interpretation as to what the systems approach is. Strength comes from the approach's being so well endowed with breadth which means not only that it has usefulness in many different contexts, but also, through vagueness, it maintains a kind of featherbed resilience against attack and hence a notable vunerability against criticism. But lack of articulation conveys weakness too, especially as the approach usually carries the tag scientific and precise. Hoos (1972) identified three root causes of the difficulties, contradictions, and complexities of the systems approach. The first one is the looseness of the word 'system', each discipline having its own idea of what a system is, along with its own definitions, principles, assumptions, and hypotheses. The second root cause is laxity in the usage of terms, the phrases 'systems analysis', 'systems engineering', and 'systems managment', for instance, being virtually interchangeable. The third root cause is the coming together of many and diverse disciplines and intellectual streams that have somehow been rendered happy bedfellows through semantic similitude.

3.3. Answering the charges against systems analysis in geography

These objections to systems analysis seem quite understandable. Nevertheless, Huggett (1976) argued that the eschewing of systems analysis by the vast majority of geographers stems more from the current atmosphere of cautions empiricism than from any inherent inadequacy of the systems approach. He tried to mitigate the cons with the following pros.

Firstly, it is the very complexity and long-term changes of systems of interest to geographers which prelude the microscopic and, in part, empirical methods of study and which call for a systems attack. The macroscopic relations between system components can be fruitfully explored using a systems approach. Macro-scale theories may be formalized by constructing system models; these models are often dynamical and can be analysed, among other ways, by systems analytical techniques. Consequently, it is at this level of resolution and in this context that systems analysis is usefully applied. In a dynamical system model, emphasis is on how the system functions so that, from a knowledge of the present state of the system, it is possible to compute future and past states. The advantage of this approach is that changes in systems which occur too slowly to be observed can be studied directly and not by statistical extrapolation from measured present-day rates of change. Thus, for instance, future population changes in England could be projected using a system model of inter-regional population growth, rather than using a simple statistical curve fitted to existing population data and extrapolated into futurity.

Another advantage is that cause and effect, rather than being seen as a direct link between two system variables, is viewed as a web of complex and subtle interaction in which tampering with one cause-and-effect link may lead to unexpected effects in other parts of the system. So, in the world social system, intuitive cause-and-effect thinking might lead to the conclusion that birth control would reduce world population growth. However, the world social system is vastly complex and computer simulations have shown that, though a reduction in birth-rate will lead to an initial decline in population, the effect of less people will gradually alter other variables in the system, some of which will tend to relieve the pressures which originally initiated population control, and population growth will resume. Because the behaviour of complex systems may be contrary to that which would be suspected intuitively, it has been named 'counter-intuitive behaviour' by Forrester (1971).

Secondly, and understandably, most scientists do not feel justified in describing anything not derived from empirical evidence (Watson, 1969). But, as Kline (1973) explained, it is just because systems models compute changes over time periods and in situations where such change cannot be directly observed that it is unrealistic to require complete empirical validation: the predicted population of England in the year 2000 must wait a couple of decades before it can be directly validated. Moreover, it is possible to make partial tests on the models. For example, in some cases historical residua of past system states may be matched against computed change: the population model for England could be run with a base date of say, 1920, and the predictions (postdictions) can be matched with known changes since 1921. Hence, though it is difficult to obtain complete validation of the models, in many cases a combined programme of theoretical and empirical work should enable the validity of a system model to be tested.

Thirdly, systems analysis can deal with non-linear relationships, with variable lags and thresholds, and with conditions of disequilibrium (see Bledsoe and Van Dyne, 1971: Kowal, 1971). To model systems which lurch from one state of disequilibrium to another is admittedly a daunting prospect, but a systems attack on this disputatious topic is feasible and might stand more chance of success than a purely empirical approach, especially where the system involved is complex. Shooting and streaming flow in streams, for example, represent alternative solutions to Bernoulli's equation. That this is so that not surprising as intuitive reasoning might suggest that the two flow phenomena would obey the same law. In more complex systems, however, where the laws are little known, or unknown, it may be exceedingly difficult, even heuristically, to anticipate that two or more spatial manifestations of a geographical system are merely equally valid solutions to given input and boundary conditions. For example, in complex global climatic models it has been speculated that there may be several plausible equilibrium temperature distributions for a given set of input variables (Sellers, 1969, 1973). Similarly, Gersmehl (1976) pointed out that mineral cycles in ecosystems, where flows are non-linearly determined, may not have just one equilibrial state and this allows for several distinct vegetation climaxes at the same site; we have met this idea already in the guise of neighbourhood stability (p. 6). Theoretical analysis of geographical systems might therefore reveal such plural solutions. Furthermore, if there is more than one solution then the system may lurch from one 'solution' to another through time (as suggested by Curry, 1962 and made explicit in catastrophe theory) and these temporal differences might, at any instant, be expressed as spatial differences. That the spatial patterns concerned are in any way connected might prove most unexpected and not at all apparent from empirical work.

3.4. What can geographers do with systems analysis?

Admittedly, the kind of discussion proffered by Huggett, which serves only to extol the virtues of systems analysis in geography, is unlikely to convince the unconverted. A change of heart would more likely be engendered by some concrete examples of systems analysis bringing new light to some geographical problems. This is the nitty-gritty of the problem: we as geographers must carefully and thoughtfully work out how and why systems analysis is relevant to us, if indeed it is. In Harvey's (1969) terms, we have to interpret the calculus of systems analysis in a geographical context.

On this theme, it is interesting that the first two phases of systems analysis—the lexical and the parsing phase—loosely follow the first two of the three themes given by Chisholm (1975) as a rough-and-ready definition of geography. These themes are: (1) the recording and description of phenomena at or near the earth's surface; (2) the study of relations between phenomena at particular places; and (3) the special examination of problems in the framework of terrestrial space. The omission of the last theme from the strategy of systems analysis gives a clue as to why systems analysis has not yet been too satisfactorily applied in many branches of geography and as to what geographers need to do to make systems analysis more amenable to the study of uniquely geographical problems: perhaps to be successful in geography, systems analysis must be adapted to tackle intrinsically spatial structures. Certainly, in the fields where it was developed, systems analysis was not explicitly designed to cope with spatial problems so perhaps it is up to geographers to take the initiative in this relatively unexplored field. However, this suggestion looks over a cupboard-dwelling skeleton or two: for although spatial aspects of geography are usually taken unquestionably as the core of the subject, there are a few profane practitioners (Huggett, 1976; Eyre, 1973) who can see no case for geographers' claiming spatial study as their preserve. After all, the theoretical work of economists—von Thünen, Weber, Lösch—is the Juggernaut of many modern locational theorists; Newtonian principles concerning the motion of bodies in planetary and atomic space have been adapted by geographers to the study of movement in terrestrial space; and even the recent work on the perception of geographical space has made its inroad from psychology. None the less, though one might choose to decline to go along with Abler, Adams, and Gould (1971) and take the consideration of spatial problems as the geographer's *raison d'être*, one must admit it is now timely to examine the possibility of applying systems analysis to the growth, development, and perhaps perception of spatial structures within the geographical realm.

The rest of the book examines systems analysis in a geographical setting. The structure is as follows. The phases of systems analysis—the lexical phase, the parsing phase, and the modelling and analysis phases—are explored in turn.The modelling and analysis phases are, for discussion, difficult to separate; they are therefore combined and dealt with in two chapters. The modelling-cum-analysis chapters consider the two chief groups of geographically interesting system models, by name, flow models—models of systems in which the circulation of matter, energy, people, goods, or money is studied without explicit reference to spatial and locational system facets; and regional models—models of systems in which the locational and spatial facets of system components and flows are an integral ingredient.

4 The Lexical Phase

> Not chaos-like, together crushed and bruised,
> But, as the world harmoniously confused:
> Where order in variety we see,
> And where, though all things differ, all agree.
>
> POPE, *Windsor Forest*

THE lexical phase of systems analysis involves two steps: the identification of the basic components or entities of which a system is composed; and the definition or description of these components in terms of measurable properties or state variables. This phase is of enormous importance but it has not always been given the attention it would warrant. The process of identifying entities has been termed entitation; the process of measuring entities has been styled quantitation (Gerard, 1969). Entitation is vastly more important a stage in systems analysis than quantitation: Gerard (1969) demonstrated this point with the example of primitive astronomy where stars (basic entities) were grouped into constellations which we now know were, in reality, meaningless.

Entitation is in part influenced by the parsing phase, by the nature of the relations by which the system components are linked; for by definition, a system component would not be a system component were it not related to other components in the system. The defining of system and sub-system units requires some presumptions of relations in the system; it also rather assumes that a system has been identified, that some meaningful structure has been separated from the real world. The identification of some geographical systems, especially those in the human realm, is beset with problems and is often met with scepticism. This chapter looks at the identification of systems of interest to geographers and considers the components and state variables of the better understood ones.

Geographical systems and their component parts have at least two facets which can be separately and conjointly identified and described. First of all, geographical systems have a functional or process facet; they can be viewed as physical systems composed of interconnected storage units which are linked by flows of energy, matter, and, in some cases, information. Examples include the basin water cycle in which storage components, such as soil moisture storage and surface storage, are linked by flows of water; and an urban system in which a city is seen as a store of energy, goods, and information which it exchanges with other settlements. Secondly, geographical systems have a structural or form facet: many of them have a distinct spatial architecture. A transport network and a central place hierarchy are intrinsically spatial structures. In reality, geographical systems at once display both functional and structural facets: the structure of a central place hierarchy is created by flows of goods and money between system units but the flows are in turn influenced by the system structure. The integration of form and process, structure and function, is seen in regional systems.

4.1. The functional aspects of systems

It was suggested in an earlier chapter that the hierarchy of natural systems which form the Earth has two main branches. The first branch consists of geological systems and includes the solid globe, the outer part of which is called the lithosphere; the hydrosphere and cryosphere, which are, respectively, the layers of water and ice that partly cover the solid globe; and the atmosphere, a shell of gas which engulfs the other systems. The second branch leads to the systems of biology and sociology: cells, organisms, and societies. The global sum of the various biological systems, which extends across the land surface and throughout the oceans, is the biosphere. For the biosphere and its component biological systems, including Man, to function, four things are required: water, nutrient minerals, certain gases (carbon dioxide and oxygen), and sunlight; these four requisites are supplied to man and other organisms by their environment or surroundings, that is, the other earth systems. Thus biological systems interact with other earth systems, as well as between themselves. The resulting interacting complexes are called ecological systems and their global total the ecosphere. Ecosystems, as they are usually termed, thus consist of two co-active parts: one part of living organisms or biotic part; and one part of non-living systems or abiotic part. Interaction between biotic and abiotic systems involves a transference of matter and energy.

To analyse ecosystems it is helpful to identify and isolate the component parts of the system and to establish the dynamical processes by which the transference of matter, energy and, in socio-ecological systems, information and money is realized. Such a procedure emphasizes system function as opposed to spatial structure. Of course, system processes take place in, and are organized in, space; but in this section stress is given to flows between system units viewed in an essentially non-spatial context.

Of special interest to geographers are several, virtually endless, recirculatory processes which pump matter, energy, and information through ecosystems. Among their number is the water cycle (or, in acknowledgement to the Greek tongue, the hydrological cycle), the rock or sedimentary cycle, the nutrient or biogeochemical cycle, and the flow of goods, money, and ideas through human ecosystems. Each of these cycles can be represented as a series of storage components linked by dynamical processes which effect the transfer of water, rock, or whatever between stores. Such an arrangement, which is often portrayed by the familiar box-and-arrow diagram, constitutes a system. Indeed, the very process of identifying storage components and links between them, and sketching the same as a flow diagram is often an initial step in systems analysis. The creation of a flow diagram may seem an elementary step; but the process certainly brings to light any gaps in our knowledge of a system and can indicate areas of enquiry to which experimental work might fruitfully be directed. To illustrate this point, albeit rather crudely, it might be found that, in a flow diagram of the water cycle, little is known about the link between ground-water and streamflow; fieldwork could thus be initiated to fill this gap in knowledge.

It should be stressed here that many distinct representations of the cycles could be made according to the goals of the analyst. Thus soil scientists might focus on, and make subdivisions of, one set of components of the water cycle, meteorologists another. Similarly, glaciologists would require more detail on the glacial components of the cycle. Thus we come to a conclusion already noted (p. 16), that all systems are models of reality which stand to be refuted; it is always important to bear this point in mind.

4.1.1 *System representation*

It is helpful at this juncture, before embarking on an account of particular process systems, to discuss briefly the various ways in which systems can be depicted. Flow diagrams have already been mentioned as one way of depicting a system. At their crudest, flow diagrams consist of boxes with arrows indicating links. More sophisticated forms of flow diagram exist which, more than just indicating a link, assume something of the relations between components. Each component is seen to perform some standard function and is accordingly represented as a canonical structure (this simply means that the component is a basic functional unit, the behaviour of which need not be resolved at a lower level). Dozens of diagramming languages dealing with canonical structures have been devised. One now widely used symbolic language was first given by Forrester in 1961; this computer-simulation language was written, initially at least, to tackle industrial and urban problems, but later extended to the so-called world simulation, as for example discussed in *The Limits to Growth.* Some of the basic modules, as they are known, in the Forrester's Systems Dynamics language are shown in Figure 4.1. The main modules are: state variables, shown by valves; auxiliary variables that influence the rates of processes, drawn as circles; flows of people, goods, money, energy, and so on, represented by solid arrows; causal relationships, denoted by broken

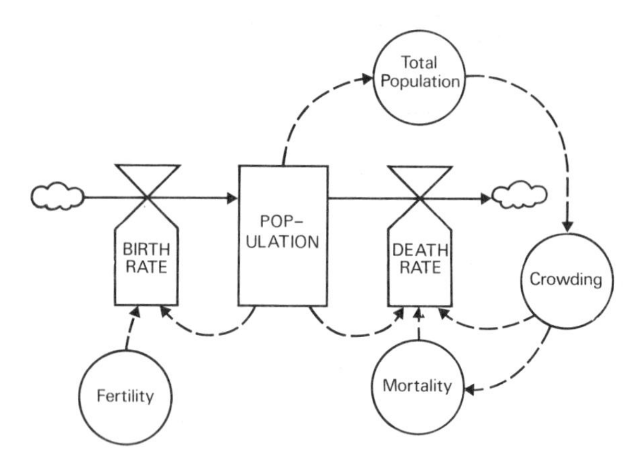

FIG. 4.1. Basic modules of Forrester's (1961) simulation language.

arrows; and clouds, depicting sources and sinks of energy and mass flows. In a model of world population, the population in various age classes would be state variables, the rate of change of which would be influenced by birth- and death-rate valves, one for each age class; total population would be an auxiliary variable causally linked to the state variables; the links between each age class would represent actual flows of people.

Another symbolic language was devised by Odum (1971). The basis of his language, known as energese, is that certain modules represent a particular structure and function within a system, and lines are pathways of energy. In isolation, the components of this energy circuit language look like the doodlings of an eccentric electrical engineer; but a concrete example, to wit, energy stores and transfers in a medieval village, should serve to show the utility of the language. A schematic diagram of the village system is shown in Figure 4.2a, and its translation into Odum's energy circuit language in Figure 4.2b. The energy flows are shown by lines and potential energy sources, such as food, by a circular symbol. Passive energy storage, such as grain in the grain store, is shown by a tank symbol. The pointed block (a sort of stubby arrow) is a work gate and represents the work done by people procuring energy from sources outside the village; it is thus a kind of control valve. The hexagon represents a self-maintaining sub-system and is a combination of two modules—a work gate and potential energy store, which act in concert to ensure that the energy stored is fed back to control the work done by the whole unit (Figure 4.2c). The arrows going to ground depict that portion of energy which is dissipated as heat while work is being done. The bullet-shaped symbol represents the reception of pure wave energy such as sound, light, water waves, and wind; an example is a windmill in which

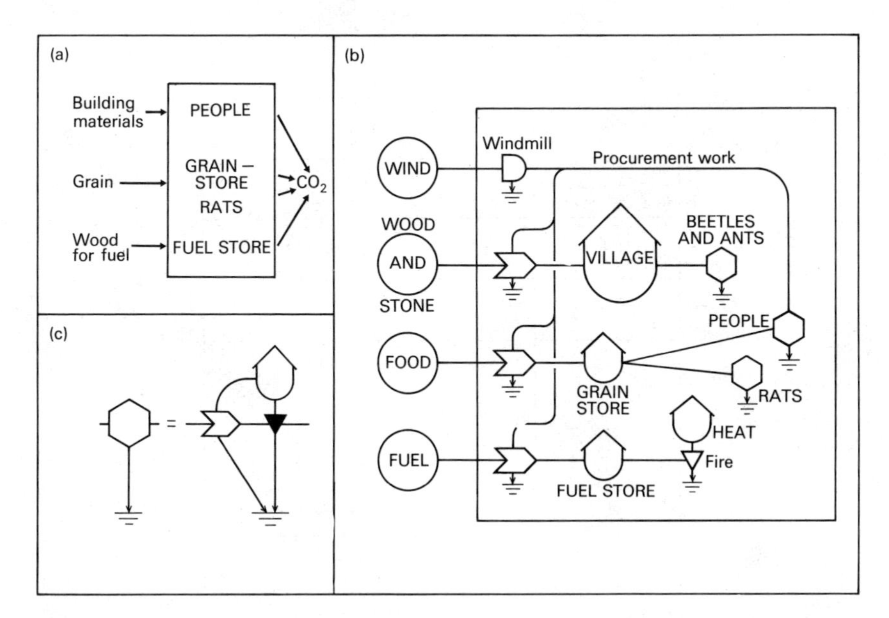

FIG. 4.2. Energy stores and transfers in a medieval village expressed in Odum's (1971) energese.

incoming energy interacts with the sails of the windmill to produce kinetic energy for grinding grain.

Yet another symbolic language, with its origins in the hydrological literature, was adopted by Chorley and Kennedy (1971) to depict physical process–response systems. The basic components of this language are shown in Figure 4.3. Severally combined, these components can represent a variety of systems. For instance, Figure 4.4a shows, in conventional diagrammatic form, the annual transfers and stores of solar energy (short-wave radiation) through the earth-atmosphere system; Figure 4.4b portrays the same information as a canonical structure. It can be seen that 263 kly of solar energy are received at the top of the atmosphere in one year. Of this total, 63 kly are reflected back into space by clouds and 15 kly by air molecules, dust, and water vapour; clouds absorb 7 kly and air absorbs 38 kly; 139 kly reach the earth's surface as direct beam and diffuse radiation, but 16 kly of this are reflected by the earth back into space leaving 124 kly absorbed by the earth's surface.

Flows between system components, apart from being shown pictorially, can also be displayed as a table or matrix. In such a matrix, row and column headings are the system components: in the solar-energy example, the headings are space, clouds, air, and land (see Table 4.1a). The elements of the matrix may simply indicate the presence or absence of a

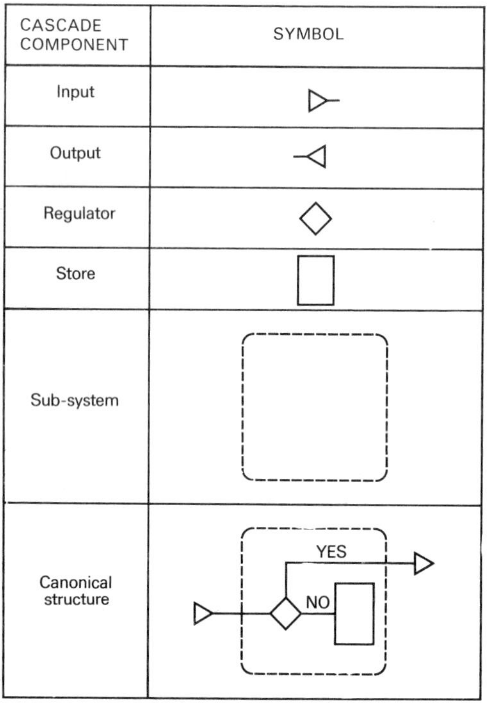

CASCADE COMPONENT	SYMBOL
Input	
Output	
Regulator	
Store	
Sub-system	
Canonical structure	YES NO

FIG. 4.3. Symbols used by Chorley and Kennedy to depict canonical structures. Reprinted with permission, from R. J. Chorley and B. A. Kennedy (1971).

link between components by a 0 or 1. The links in the solar-energy example are shown in this manner in Table 4.1a. Thus, the 1 in position row 1 column 2 indicates that there is an unspecified transfer of short-wave radiation from clouds to space. The matrix elements may also be the actual flows of energy between components; these flows are shown for the solar-energy case in Table 4.1b. In Table 4.1c, the flows represented by the matrix elements are indicated by the same symbols as in Figure 4.4a. Although this example of expressing system flow processes in matrix form is rather trivial, the procedure can form the basis of sophisticated forms of systems analysis, such as input–output flow analysis of ecological and economic systems (this will be dealt with later in the book).

4.1.2. *The water cycle*

Of all phases of the global water cycle, the transfer of water off the land towards the oceans, the land phase, is of special interest to the geomorphologist. The movement of water over the land is organized as a distinct run-off system consisting of two interrelated parts; the one part being a

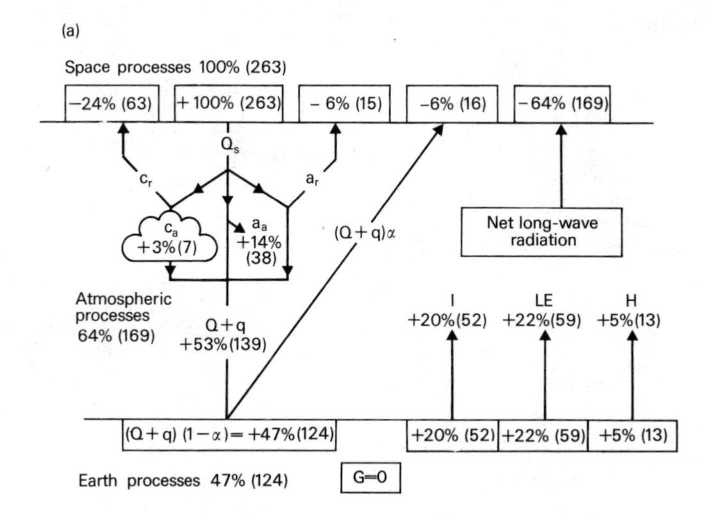

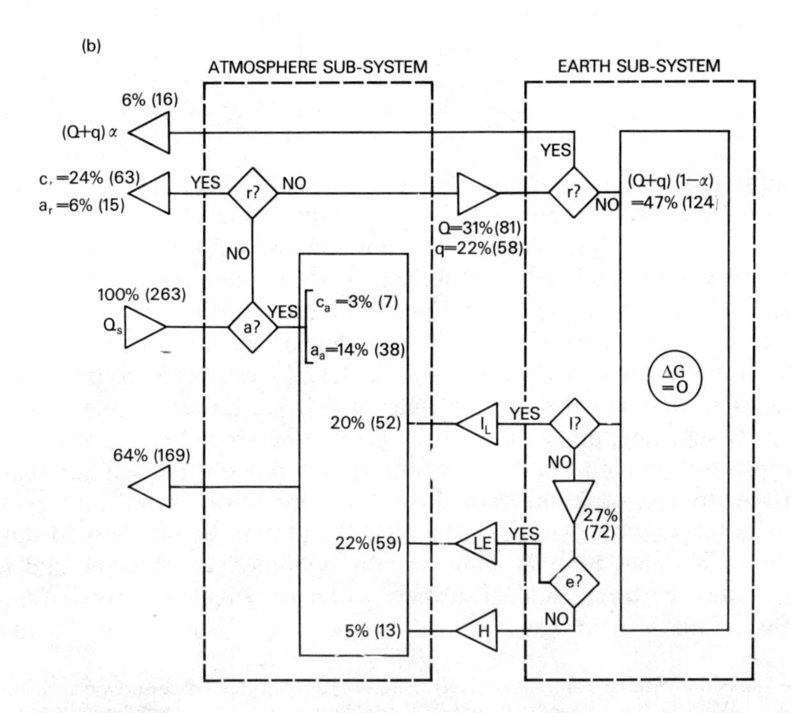

FIG. 4.4. Transfers and stores of solar energy in the earth-atmosphere system: (a) conventional flow diagram; (b) canonical structure. Reprinted with permission, from R. J. Chorley and B. A. Kennedy (1971).

Table 4.1

(a)

		from space	clouds	air	land
to	space	—	1	1	1
	clouds	1	—	0	0
	air	1	0	—	0
	land	1	0	0	—

(b)

		from space	clouds	air	land
to	space	—	63	15	16
	clouds	7	—	0	0
	air	38	0	—	0
	land	124	0		—

(c)

		from space	clouds	air	land
to	space	—	Cr	Ar	$Q(1-\alpha)$
	clouds	Ca	—	0	0
	air	Aa	0	—	0
	land	$(Q+q)(1-\alpha)$	0	0	—

surface drainage system; the other part being a sub-surface drainage system. The flows of water in both these systems take place within distinct areal units of the land surface. The flows in the surface drainage system take place within what the American literature refers to as a drainage basin or watershed and what English literature usually refers to as a catchment.[1] The flows of water in a catchment are said to constitute the basin water system as shown in Figure 4.5. As can be seen, this system consists of a series of stores of water which are linked to one another. Broadly speaking, precipitation that reaches the ground enters the soil by the process of infiltration, is temporarily stored on the soil surface as surface storage, or is intercepted by, and temporarily stored on, vegetation surfaces. Water stored in or on the soil may be transmitted down valley-side slopes by overland flow, throughflow, and pipeflow and fed into rivers to form channel storage, or may percolate downwards to recharge the ground water store. Ground waters may rise by capillary

[1]The flows in the surface and sub-surface drainage systems will not necessarily be coincident in plan. Whereas the catchment boundaries are determined by topography, the boundaries of the sub-surface system are determined by the disposition of the water table or phreatic divide. Eagleson (1970) suggested that the term 'catchment' should be confined to the surface system and the drainage basin to the combined surface and sub-surface systems.

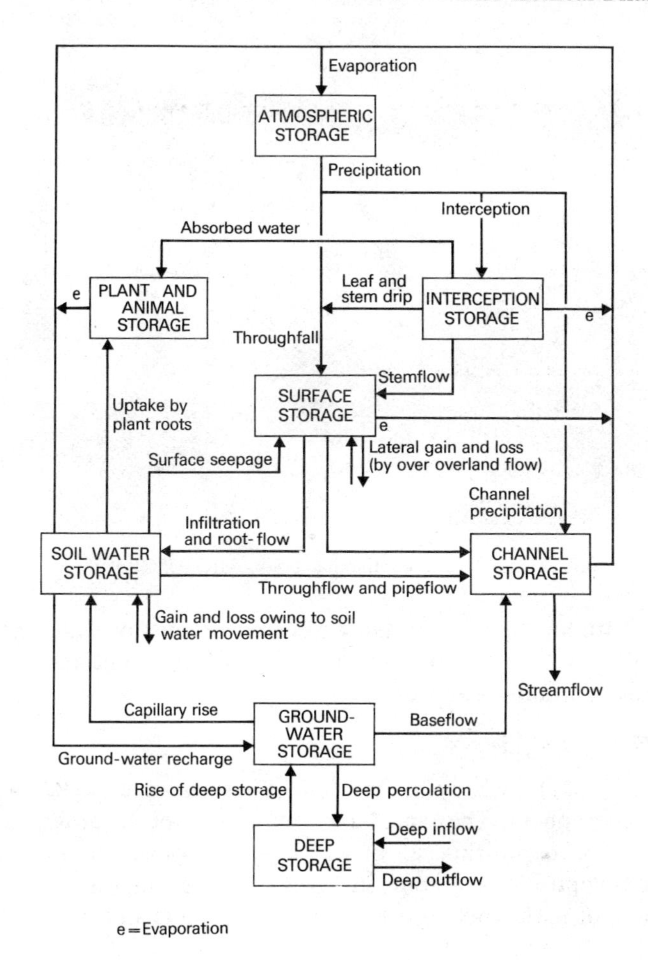

FIG. 4.5. The catchment water system.

action and pass in to soil storage. Water in any of the stores may be evaporated back to the atmosphere. As we shall see later in the book, this broad conception of the water cycle forms the basis of many systems models which are used, for example, to predict floods.

The state of a hydrological system is described by the amounts of water in each of the component stores at a particular time. The amount of stored water may be expressed in a variety of units. Storages in the global hydrological system are expressed by Nace (1969) as volume of water (thousands of km^3). Assuming that the density of water does not vary significantly from 1 g/cc, these figures could be readily converted to mass of water in store. The transfers of water in the hydrological cycle can be expressed as volume or mass per unit time but, especially in global and

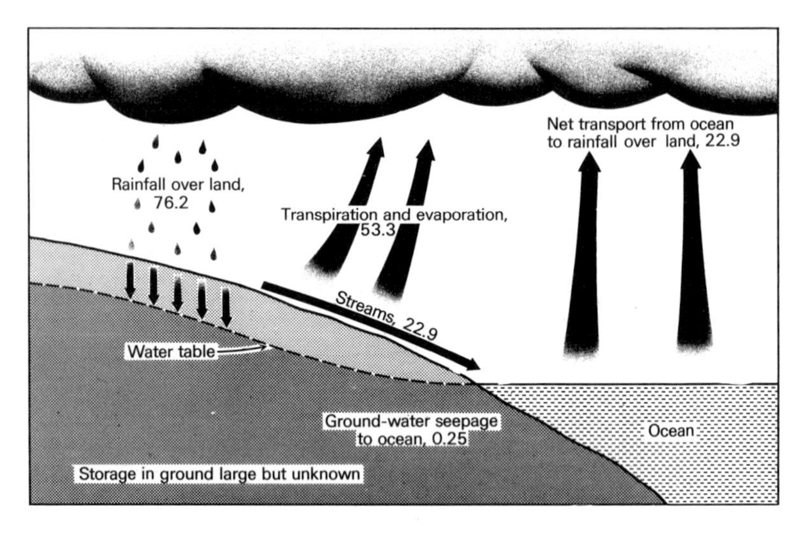

FIG. 4.6. The water budget of the continental USA. After USGS.

continental studies, they are often expressed in units of water equivalent depth: Figure 4.6 shows the water budget of the continental USA in centimetres of rainfall equivalent.

4.1.3. *The rock cycle*

Within the global rock cycle, the land phase is of especial interest to physical geographers. The land phase comprises the several processes of denudation and deposition, all of which can be viewed as a system. The denudation–deposition system involves storage, inputs, outputs, and throughputs of rock, rock debris, and other products of weathering such as colloids and solutes.

A basic organizational unit of denudation–deposition processes is the catchment. The chief processes in the catchment are as follows (Figure 4.6). As the depth of weathering increases, usually in step with the erosional lowering of the land surface, so fresh rock is incorporated in the system. Material may be added to the upper boundary of the system, the land surface, by deposition, having been transported by wind and ice, not a few kilometres or less, for that constitutes local movement within the system, but across and between continents. All the materials in the denudation–deposition system, both local and far travelled, are subject to transformation by the complex processes of weathering. Some weathering products, by further transformations which include the production by organisms of resistant organic and inorganic materials to give their bodies structural support or protection, revert to a rock-like state. The weathered mantle tends to move through the catchment: a throughput of

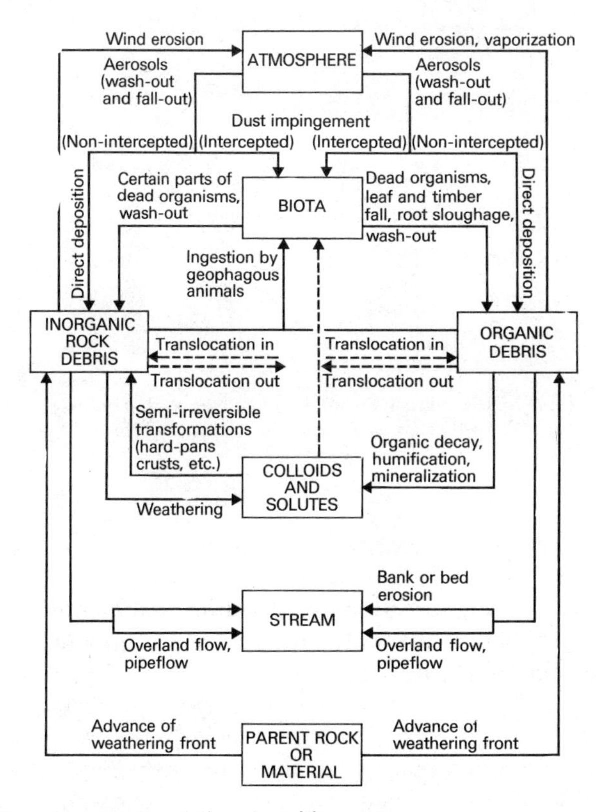

FIG. 4.7. The catchment denudation–deposition system.

weathering products is effected by mass movements, surface wash, rill action, gullying, and bank and bed erosion in streams. Wind does remove material from the system but the greater loss is carried in rivers to the oceans.

The state of the denudation–deposition system would ideally be expressed as the mass of materials in the different stores. In fact the store sizes are very difficult to estimate except in gross terms (see, for instance, Stoddart, 1969). Emphasis is usually given to morphological descriptors of the denudation–deposition system. Four morphologically distinct systems have been recognized in the landscape: the slope system, which may be subdivided into slopes of interfluve areas, slopes of valley sides, and sea-cliffs; the river-channel system; and the drainage-basin system (Chorley and Kennedy, 1971). Descriptors of these systems are legion. For instance, geometrical properties of interfluvial-slope systems include angle of slope, the depth of weathered mantle, the principal grain-size characteristics of the soil and mantle, the acidity of the soil and mantle,

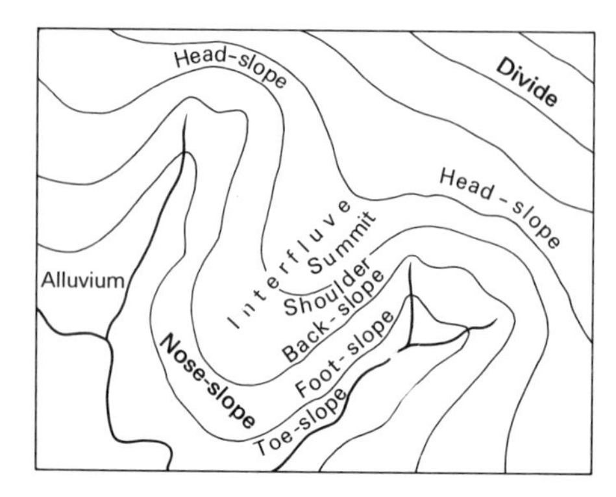

FIG. 4.8. Geomorphological and slope profile components of a drainage basin. The terminology follows that given by Ruhe and Walker (1968).

the soil wetness, the root weight of the soil, and so forth. And, the principal variables in the river-channel system, according to Lane (1957), are discharge, sediment load, velocity, bed roughness, nature of bed and banks, bed slope, vegetation, water temperature, and human interference.

The division of the land surface into four component systems is not the only possible scheme. At a small scale, for example, McPherson (1969) argued that it is better to study the river system rather than just the river-channel system, the full river system including not only channel variables but also flood-plain characteristics such as channel pattern, channel shift, flood-plain height, morphology, stratigraphy, and vegetation. At a larger scale, Ruhe and Walker (1968) recognized a different set of slope types within an entire drainage basin, viz, divide, interfluve, nose-slope, head-slope, and alluvial fill (Figure 4.9). Within head-slopes and side-slopes they defined five slope-profile components—summit, shoulder, backslope, footslope, and toeslope—for an open system in which the drainage basin is part of a more extensive drainage network; and three slope components—divide, peripheral slope, and depository—for a closed basin, that is, an area of inland drainage. In arid environments, other descriptors of slope have been used. Cooke (1970) studied the gross morphometric properties of pediments and their associated landforms in the western Mohave desert, California. In this work he firstly set up working definitions of the landforms he studied within fifty-three pediment associations (Figure 4.9a). A pediment association includes the pediment, the mountain area tributary to the pediment, and the area of alluvial plain to which the pediment is tributary; it is thus a complete denudation–deposition system. Secondly, he used several geometrical descriptors of the landforms, most of them pertaining to length, height, and slope (Figure 4.9b).

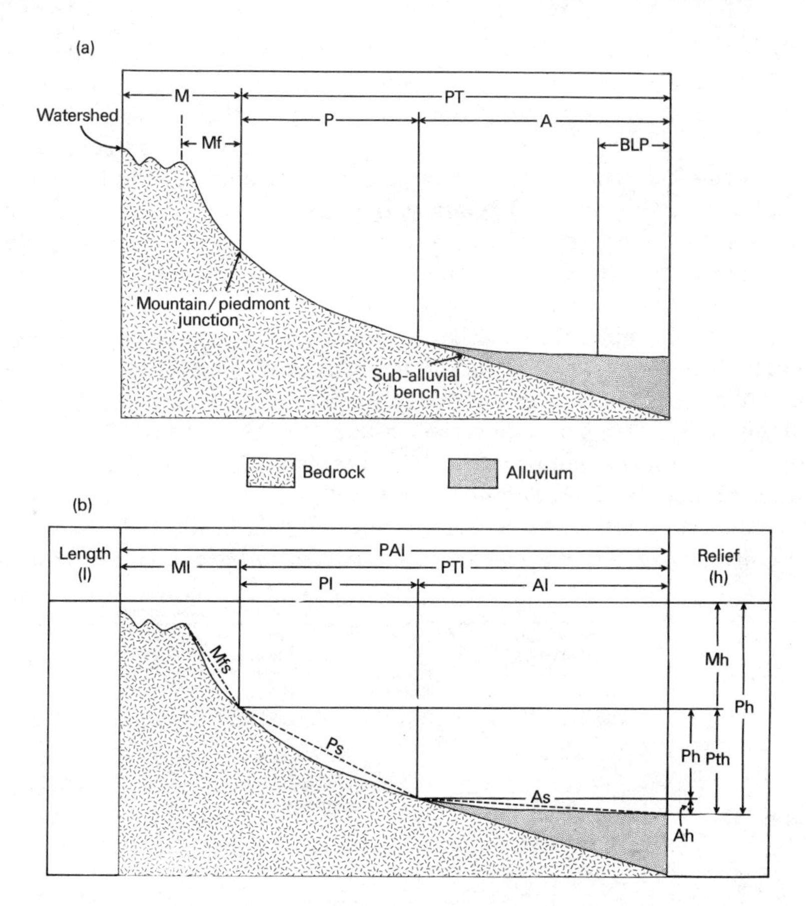

FIG. 4.9. Landforms and their description in the Mohave desert. Key to notation: *Landforms*—M = mountain area; Mf = mountain front; PT = piedmont plain; P = pediment; A = alluvial plain; BLP = base level plain; PA = pediment association: *Properties*—1 = length; h = relief; s = slope. Reprinted with permission, from R. U. Cooke (1970), *American Journal of Science*, **269.** 26–38.

4.1.4. *The biogeochemical cycle*

The biosphere powers a global cycle of carbon, oxygen, hydrogen, nitrogen, and other mineral elements. These minerals circulate within the biosphere and between the biosphere and its environment. The roughly circular paths of minerals are called biogeochemical cycles, the cycles of those minerals deemed essential to life being referred to as nutrient cycles. Two basic types of biogeochemical cycle may be distinguished. Firstly, gaseous types, in which elements such as oxygen and carbon dioxide are transferred in the gaseous state between the biosphere and atmosphere or oceans. Secondly, sedimentary types, in which matter, such as cations of calcium and magnesium, is transferred in solution between the biosphere and hydrosphere or lithosphere. Many of the processes in the land phase of sedimentary-type biogeochemical cycles are organized in the framework of the catchment and it is possible to construct a biogeochemical system. The storage components and flows in this system are shown in Figure 4.10. Only two minerals are exchanged directly between the atmosphere and organisms, viz, carbon dioxide and oxygen; the remainder, including nitrogen which is the predominant constituent of the atmosphere, have to enter via the soil. Plants and other organisms require certain minerals which are won with great difficulty and in small amounts from unweathered rock. The transfer of mineral elements from rocks to plants is facilitated enormously by the soil, and especially the soil colloids, which forms a big reservoir of ecosystem nutrients. This mineral store in the soil is maintained by rock weathering, by the decay of organic matter (mineralization), and by any mineral inputs that might accompany precipitation. Plants draw on the nutrient reserves in the soil; the minerals taken up in this wise are temporarily immobilized in the biomass

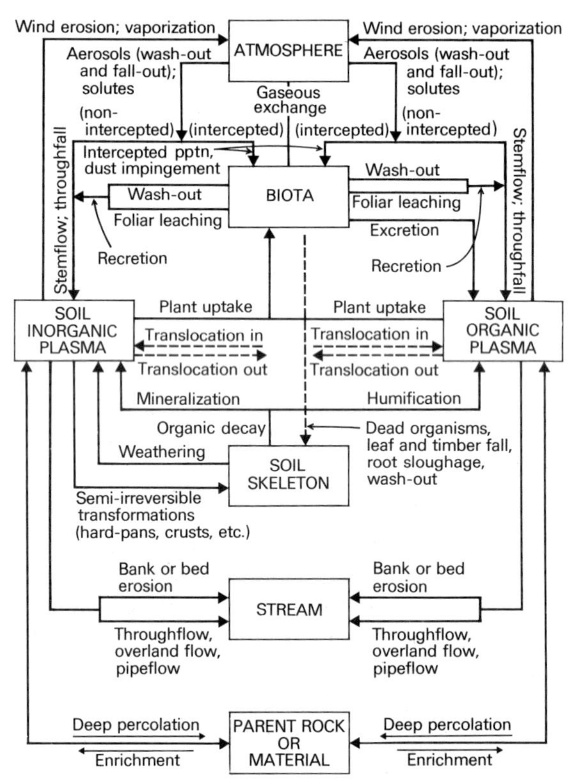

FIG. 4.10. The catchment biogeochemical system.

(living material) of ecosystems, but, in due season, are returned to the soil via foliar leaching (the washing of root-absorbed elements from leaves), litterfall and timberfall (littering), root sloughage, excretion, exudation, and the demise of organisms. The nutrients in the basin may be depleted by fire, which causes vaporization and hence release to the atmosphere of many minerals; by wind erosion; by transformations to unavailable forms; and by loss to streams via throughflow, pipeflow, overland flow, and stream bank and bed erosion. In some basins, minerals are exchanged with ground-water storage.

The state of a system with regard to minerals circulating in it is described by the quantity or concentration of a mineral in each of the stores, sometimes called pools, of which the system is composed. In Figure 4.11, the pool sizes of nitrogen in an undisturbed northern hardwood forest ecosystem at Hubbard Brook, New Hampshire, as well as the accretion rates in the pools, the bracketed figures, and nitrogen

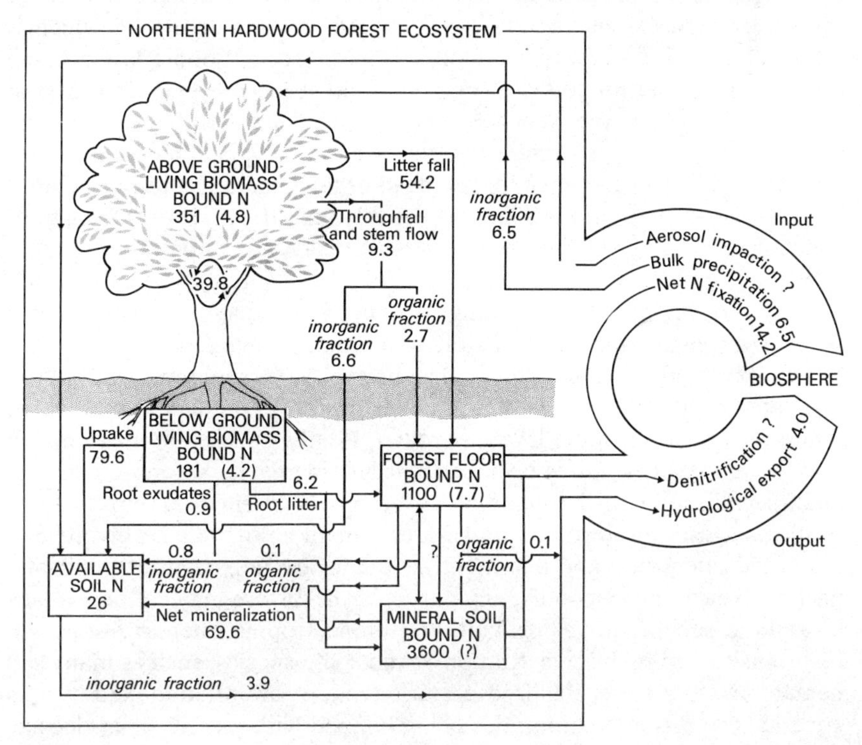

FIG. 4.11. The nitrogen budget in a northern hardwood forest at Hubbard Brook, New Hampshire. Reprinted with permission, from F. H. Bormann *et al.* (1977), 'Nitrogen budget for an aggrading northern hardwood forest ecosystem', *Science*, **196.** pp. 981–3, Fig. 1 (27 May 1977). Copyright 1977 by the American Association for the Advancement of Science.

transfer rates are shown. The pool sizes in this case are expressed in units of kilogrammes of nitrogen per hectare and all the transfer and accretion rates in units of kilogrammes of nitrogen per hectare per year; in shorthand these would be expressed as kgN/ha and kgN/ha/yr respectively, the N signifying that the mass of nitrogen is being referred to. In like manner, gC/ha would be grammes of carbon per hectare in the carbon cycle. In some mineral cycles it is necessary to partition stores of a mineral into, say, assimilated and labile storage or organic and inorganic phases (as is done in Figure 4.11).

4.1.5. *Ecological systems*

Ecosystems may be divided in to units of organisms with like feeding habits, known as trophic levels, which are linked by flows of energy (in the form of chemical energy stored in organic matter). Two fairly distinct, mutually dependent pathways of energy can be traced in most ecosystems. The first, the grazing food web, starts with the manufacture of organic material in plants and other autotrophs, the energy of which is then utilized by a succession of organisms in other trophic levels being transferred by the processes of grazing, browsing, and predation (Figure 4.12). The second, the decomposer or detrital food web, starts in the energy stored in dead organic matter which is tapped by saprophytes (plants, notably fungi, which live on dead organic matter) and thence used by a succession of organisms, collectively called saprovores, some of which graze on the saprophytes, others of which are carnivorous (Figure 4.12).

Problems arise in establishing trophic levels as sub-system units. Trophic levels are applicable mainly to animals. Plants tend to be lumped as primary producers though sub-groups such as xerophytes, halophytes, and saprophytes could be used. Williamson (1972) indicated some other limitations of the trophic level approach. Firstly, species often do not fit into the trophic categories because their feeding habits are more complex and specialized than designations as herbivore, carnivore, and so forth would suggest; for instance omnivores can fit into both herbivore and carnivore categories, and insectivorous plants are in part autotrophic and part heterotrophic. Secondly, any particular herbivore may eat only a part of a plant, such as leaves or fruit, and broad trophic categorizations will be insensitive to such specializations. And some species, such as many soil nematodes, that are specialized bacteria feeders, are often looked over in general accounts of communities being lumped with carnivores or decomposers. Thirdly, some species, such as frogs, change their feeding habits with age.

The state of an ecosystem may be described by the amount of matter, or the energy equivalent thereof, in each trophic level; in other words, the

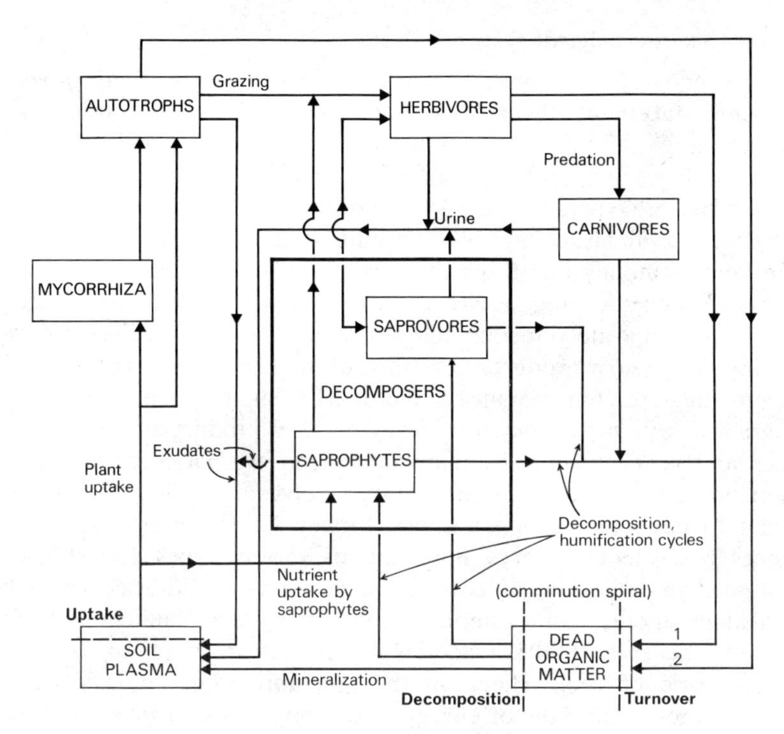

FIG. 4.12. The structure of a terrestrial ecosystem.

standing crop of each level. Units usually employed for this purpose are kilocalories per unit area, say per hectare (kcal/ha) or per square metre (kcal/m^2); flow between trophic levels would then be expressed as kilocalories per unit area per unit time, say kilocalories per square metre per day (kcal/m^2/day). Pool sizes (standing crops) are usually expressed as mass per unit area and flows between pools as mass per unit area per unit time, say grammes per square metre per year, when considering litter production and decomposition in terrestrial ecosystems. One study for instance (Reichle *et al.*, 1973), which considered the functional role of earthworms in forest litter decomposition, uses a system consisting of six state variables—rapidly decomposing litter, intermediate decomposing litter, slowly decomposing litter, soil, organic matter, and earthworms—the storage in each being measured as grammes dry weight per square metre.

4.1.6. *Socio-ecological systems*

Human geographers are interested in social systems and especially economic facets of them in relation to spatial organization. Many similarities exist between ecosystems and human economies, including currency flow and mineral cycles (Boulding, 1962). Indeed, human ecosystems are characterized by a flow of money (or its equivalent) between individuals, or higher order units such as companies and nations. The flow of money is a transaction between producer and consumer in exchange for goods and services. Money flows through a human ecosystem in the opposite direction to energy: a consumer receives goods by passing money to a producer. The flow of money can be represented as a system diagram. For instance, Figure 4.13, shows an energy–currency diagram of a human ecosystem including food production, industry, and cities and government. The diamond-shaped symbol represents a payment operation which, as is indicated, involves a small drain on energy owing to negotiations or transport. Money flows can be converted to energy equivalents. Energy flows are of various types and should be expressed in equivalent units of the same type—say kilocalories of coal equivalent energy; for example, using this system, sunlight is divided by 2000, because 2000 kilocalories of sunlight are required to produce a kilocalorie of coal. Then, on the assumption that money value is equal to the counterflow of energy, it can be expressed in units of equal quality: in 1975, 20000 kilocalories of coal equivalent was used per dollar spent.

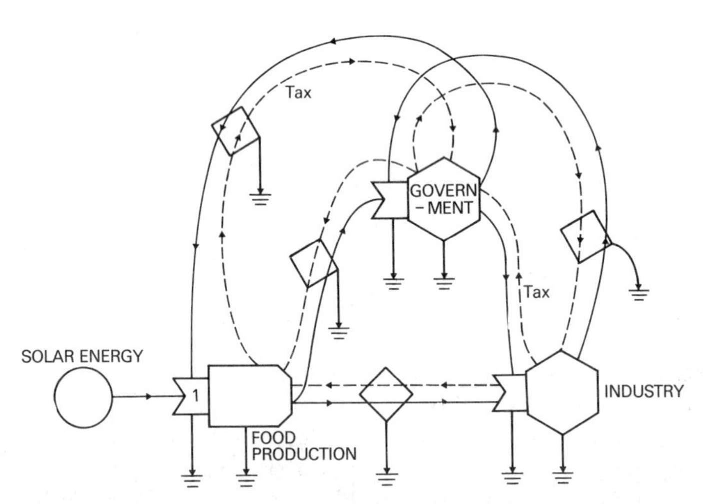

FIG. 4.13. An energy-currency diagram of a human ecosystem. Reprinted by permission, from *Environment, power and society* by H. T. Odum (1971). Copyright © 1971, by John Wiley & Sons, Inc.

As an example of an actual human ecosystem considered essentially as a flow network, The Roman Empire will be briefly discussed. The Roman Empire, perhaps the greatest world order ever built on solar energy alone, has been represented in outline by Odum (1971, p. 233) as a series of compartments linked by flows of energy and, in part, money (Figure 4.14). The Roman Government needed organizational work to develop both service provinces, such as Britain, and grain provinces, such as Africa. The service provinces provided slaves, legion recruits, special manufactured goods, and literary services; and grain provinces gave grain in the grain levy. Commerce between provinces used the currency system. Surrounding, hostile provinces had potentially competing governments and sometimes threatened to drain energy from the Roman provinces. To prevent this, the Roman legions expended a considerable amount of war

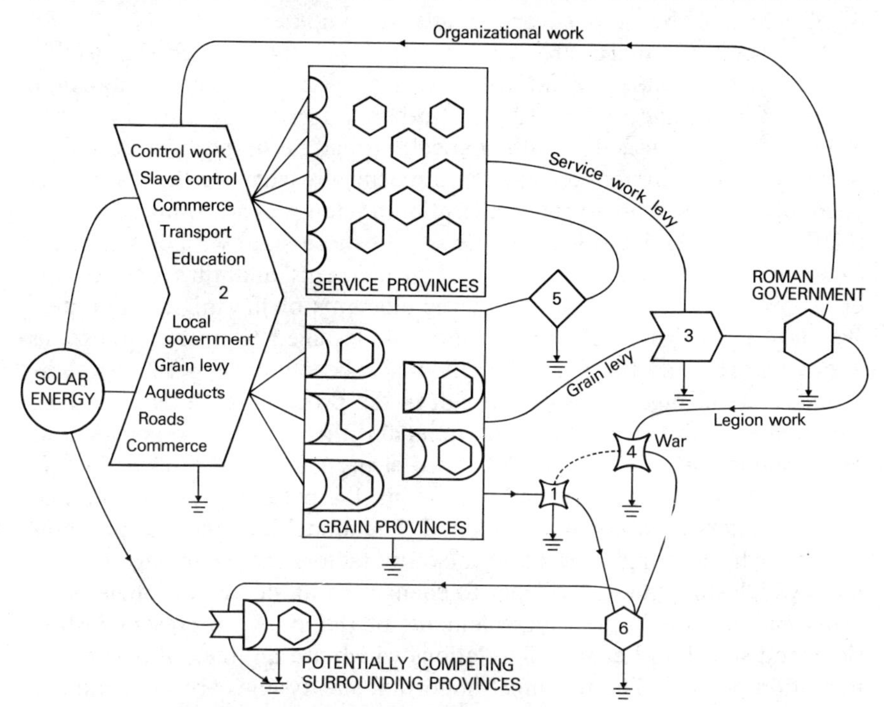

FIG. 4.14. The Roman Empire as a flow system. (1) With spread of information on organization and military methods, competing organizer countries begin to draw from some of the input energy flows; (2) developing controls; (3) service work of the provinces supplies slaves, legion recruits, and special manufactured goods; (4) military work by Roman legions prevents system drains into surrounding alternative systems by blocking the work of group 6 at switch 4; (5) intra-provincial commerce uses the currency system; (6) surrounding provinces have potentially competing governments. Reprinted by permission, from *Environment, power and society* by H. T. Odum (1971). Copyright © 1971, by John Wiley & Sons, Inc.

work. This formulation can give pointers to that unresolved problem—why did the Roman Empire decline? Looking at Figure 4.14, one possibility is a reduction in African grain production, resulting perhaps from a climatic change, may have reduced the amount of energy the government could use on organizational work to below a critical value. Though such qualitative speculation can be thought to have no advantage over other work on the Roman Empire, the point is that this human ecosystem and many others, despite their complexities of flows, can be represented as a system and, so long as the flows and storages can be measured, simulated on an analog or digital computer; this we shall consider later in the book. But as a taste of what such models are like, look at Figure 4.15a which is a systems representation of a steady-state, Vietnam-type war. The adaption of Vietnam to war is shown in Figure 4.15b which is the output from an analog computer.

Many socio-ecological processes which involve flows of goods, people, or information can be dealt with using a systems format. Mabogunje (1970), for instance, couched rural-urban migration in this way. The approach he used was not mathematical; instead he simply considered the way in which the system operates and specifically what are the interacting elements of the rural–urban migration system? And what are their relationships? Having done this, he sought those state variables external to the system which affect the system elements and thus defined the environment of the system. The basic elements of the migration system are shown in Figure 4.16. This model is designed to answer questions about the reasons for a rural individual's becoming a permanent city dweller; the changes he goes through in the process; the effects of these changes on both the rural area he quits and the city area he goes to; the encouraging and discouraging influence of particular situations or institutions on the rate of rural–urban movement; the pattern of the movements and the factors by which it is fashioned. The model identifies the potential migrant who is being tempted to move by factors in a changing environment which stimulate the villager to change the locale and rationale of his economic activities. The other elements of the system represent institutions and social and economic relations which are an integral part of the migration process. The two most important sub-systems are the rural and urban control systems; these oversee the operation of the whole system and influence the timing and amount of the rural–urban flows.

Of all socio-ecological systems, agricultural ones are the most tied to the physical environment. Agricultural systems, or agro-ecosystems as they have been called in a recently established journal of that name, in addition to the circulation of material and energy, contain man-manipulated processes which modify inputs, outputs, and rates of transfer in an ecosystem. Human intervention leads to agro-ecosystems' being

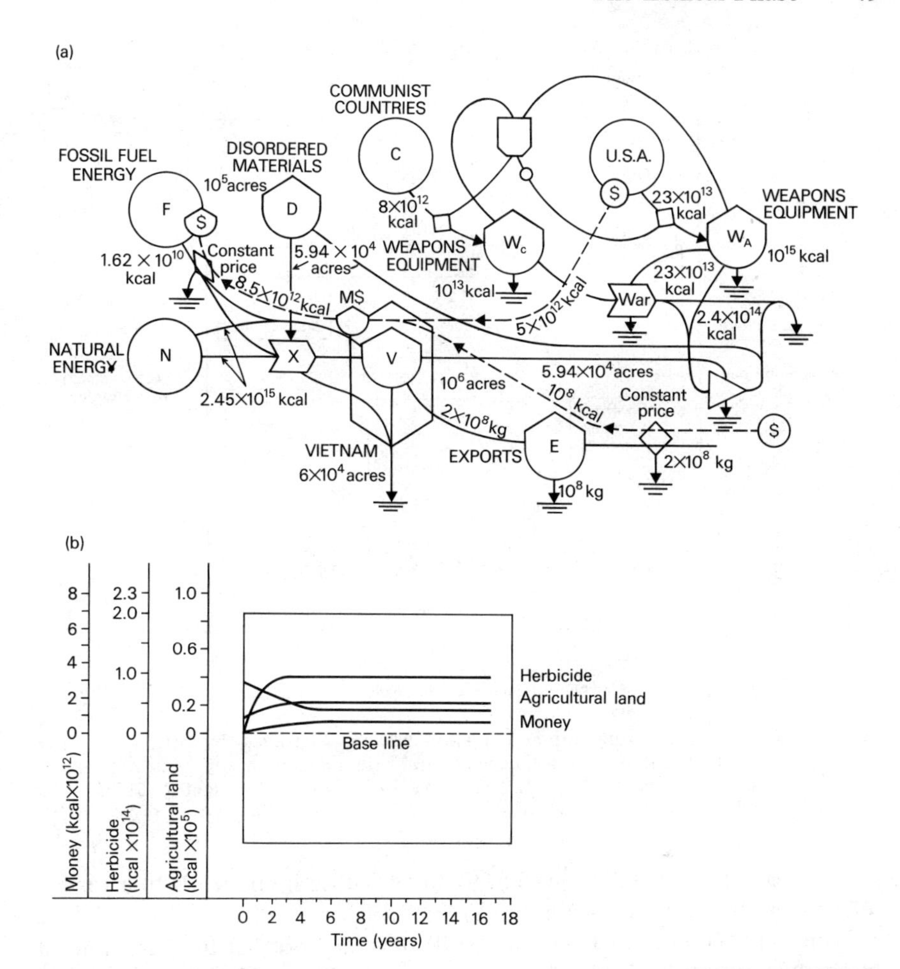

FIG. 4.15. A Vietnam-type war: (a) a systems representation; (b) a computer output showing changes in state variables with time. Reprinted with permission, from H. T. Odum (1976), Macroscopic minimodels of man and nature. In B. C. Patten (editor), *Systems analysis and simulation in ecology*, iv. 249–80. Academic Press, New York.

different from other terrestrial ecosystems in the following ways (Loucks, 1977): maximum harvest is a dominant goal achieved through the use of monocultures and of opportunistic species; the system is supported by external perturbations, and additions of minerals and water; the residual detritus of the system is reduced by harvest, export, exposure, and the continual disturbance of the soil; the majority of materials in excess of the needs of plants are readily leached into ground water or a near-by stream where they increase the productivity of aquatic ecosystems. Thus the

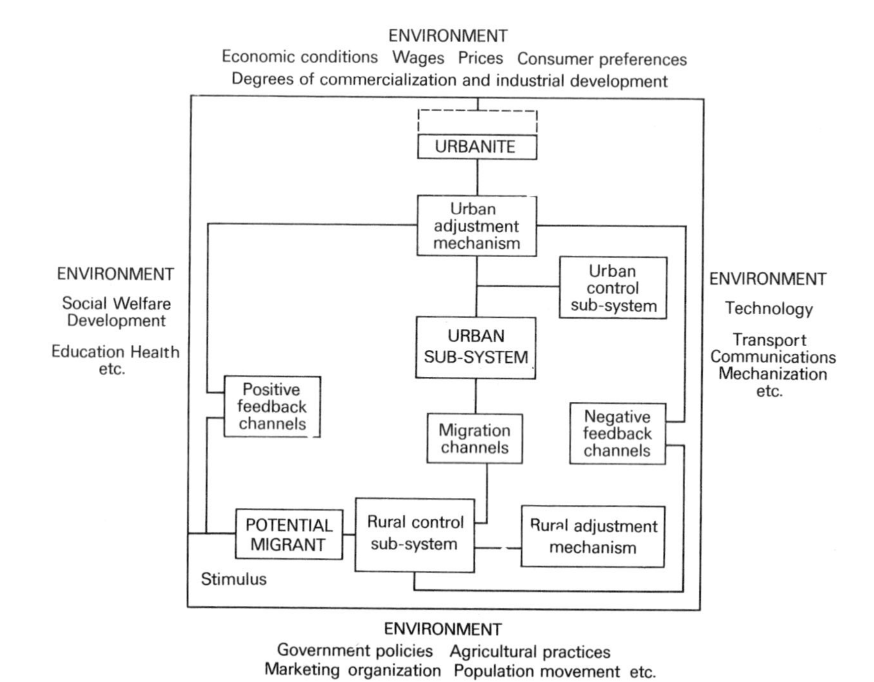

FIG. 4.16. An African rural–urban migration system. Reprinted by permission, from 'Systems approach to a theory of rural–urban migration', by A. L. Mabogunje, *Geographical Analysis*, **2. 1** (January 1970), 1–18. Copyright © 1970 by the Ohio State University Press.

conversion of a natural ecosystem to an agricultural one leads to a variety of changes in nutrient cycling and mineral flow.

Munton (1969) and Chapman (1974) have argued that the basic unit of agricultural systems is the farm system. As with agricultural systems in general, farm systems may be viewed in two ways. On the one hand, they are ecological systems with components linked by flows of energy and matter. On the other hand, farms are in part economic systems through which move money, materials, and information. The farmer who runs a farm system is thus confronted with both a physical and an economic environment; about these environments he has an incomplete knowledge, but nevertheless he has to respond to them—he has to make decisions. In this sense the farm is a transforming system in which management decisions reflect the farmer's judgement of the relations, or the relations which satisfy him, between the farm's ecological and economic constraints.

The dual facets of agricultural and farm systems can be given common expression in an input–output, flow diagram form. Figure 4.17 shows an

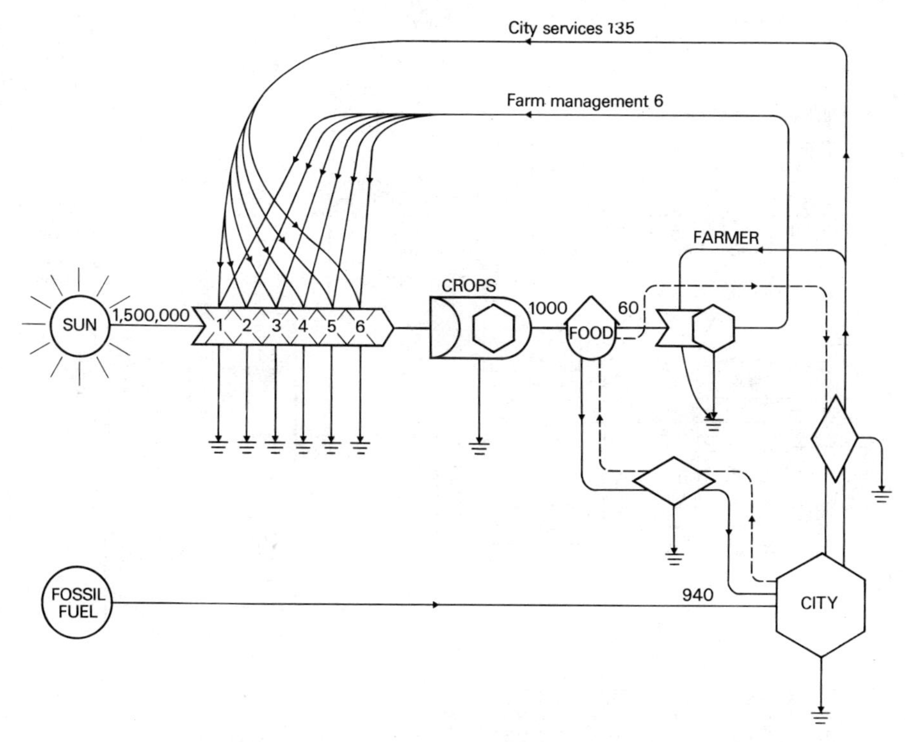

FIG. 4.17. Systems representation of an industrialized, high-yield agricultural system. Works flows include: (1) mechanized and commercial preparation of seeding and planting; (2) fertilizer additions, (3) chemical and power weeding; (4) soil preparation and treatment; (5) insecticides; (6) better varieties of crops. Reprinted by permission, from *Environment, power, and society* by H. T. Odum (1971). Copyright © 1971, by John Wiley & Sons, Inc.

industrialized, high-yield agricultural system expressed in Odum's energese (p. 32); Figure 4.18 shows a composite farm system as might be found in Britain. Of course, farming systems come in a rich variety of forms. Spedding (1975) has shown that the major types—tree crops, tillage–livestock, alternating tillage with grass, bush or forest, and grassland or grazing, any of which can be practised at varying levels of intensity—each consists of a number of different production systems which are describable in terms of sub-system units such as crop- or forage-growing system, animal-feeding system, animal-housing system, and pest- and disease-control system. In turn, production sub-systems consist of various components: different crop-growing systems, for instance, are characterized by different combinations of crop, cultivation method, fertilizer input, sowing date, and so forth.

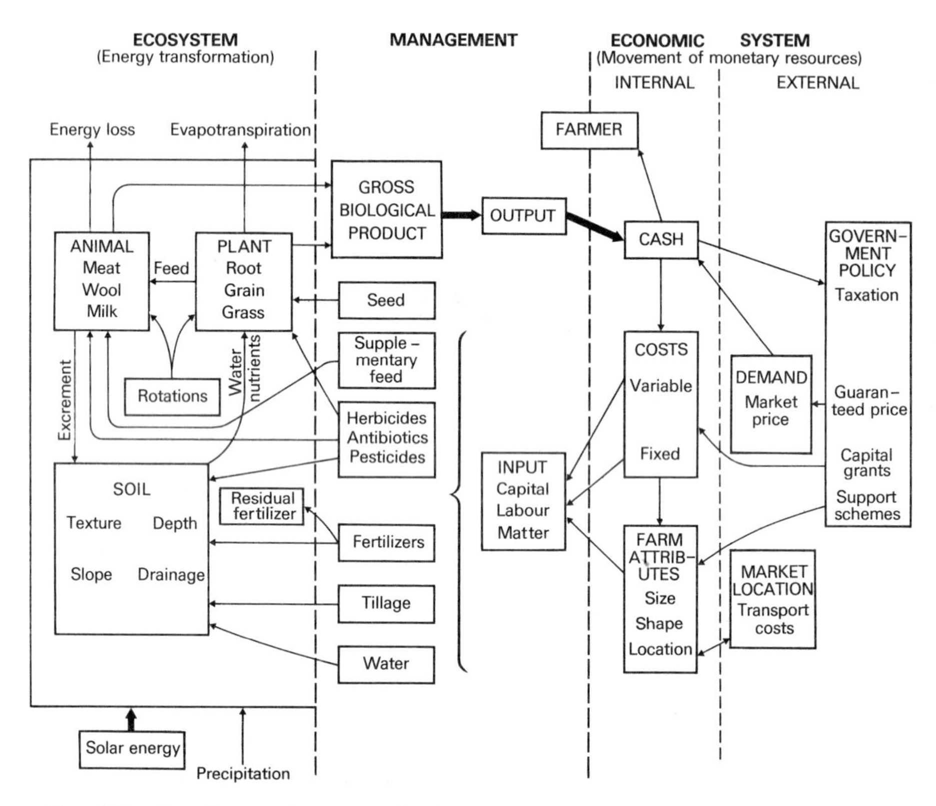

FIG. 4.18. Descriptors of a composite farm system. Reprinted with permission, from R. J. C. Munton (1969). The economic geography of agriculture. In *Trends in geography* (ed. R. U. Cooke and J. H. Johnson), chapter 14, Figure 13, p. 148.

4.2. Spatial structure of systems

Most geographical systems have an inherently spatial structure around which transactions of matter, energy, and information take place. Distinct types of spatial structure can be recognized in the geographical sphere. First of all, any system property, such as the height of the land surface or population density, may vary across space; this variation creates a surface, either real or imaginary. Secondly, surfaces are in part produced by, and in part produce, flows of matter, energy, and information across them. Some flows, like sheet-wash on slopes, may be unconcentrated; but other flows are concentrated by the configuration of the surface and thus take place in distinct channels. The channels usually form a network of paths, as in the case of rivers, roads, and communication lines, along which energy, material, and information transactions take place. In many systems, especially those of interest to human geographers, transmission, reception, and storage points of energy, matter, and information lie at the intersections of component links in network; the spatial arrangements of these points or nodes, as they are called, is a third facet of study in the architecture of geographical space.

4.2.1. *Surfaces*

The notion of a surface is easy to grasp and familiar to geographers. As a basic spatial-system component, a surface is important in creating a potential to drive a system process. It is well known that the land surface, by virtue of its height above sea level, possesses potential energy. The difference in height between two adjacent points on the land surface creates a possibility of movement of sediment and water on that surface. If sediment is moved, then some of the potential energy is used (the land surface is lowered a little) being converted into kinetic energy of the moving sediment. Generally, the rate of movement will be proportional to the negative potential energy gradient. If we call the height of the land surface h and distance x then, with the gradient of the land surface, $\Delta h/\Delta x$, equal to the potential energy gradient and with velocity v we may write (Figure 4.19)

$$v \propto -\frac{\Delta h}{\Delta x}$$

where $\propto$ means 'is proportional to'.

Traced across a surface, the direction of the potential energy gradient forms streamlines which, other factors being constant, lie at right angles to contours. On the land surface, if streamlines are traced back to points of zero potential energy gradient (that is, on watersheds) the boundaries of catchments can be defined. The boundaries of the sub-surface drainage system are defined by phreatic divides, that is, points on the water table where the gradient of hydraulic head is zero.

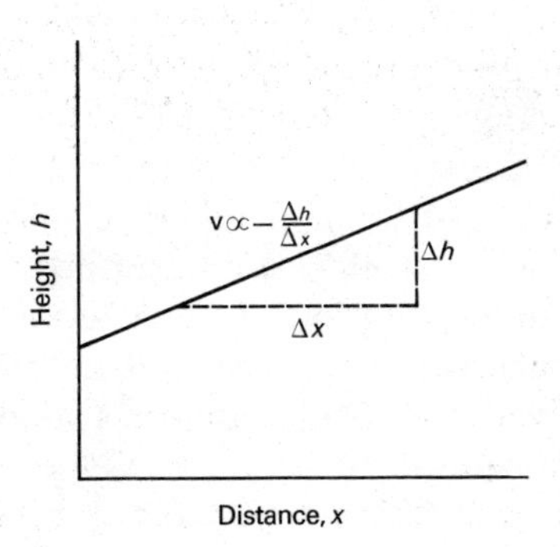

FIG. 4.19. The potential on a surface.

As with the land surface, so with the surfaces in human geography: trans-surface movements may be generated by a potential gradient. Unlike the notion of a land surface and the movement of material across it however, the notion of a surface in human geography, yet alone the generation of movement on it, is far less tangible. True, it is not difficult to see that, say, variations in population density around a city could be represented as a contour map and thought of as a surface. But can the gradient of population density be used, as the land-surface gradient can, as a measure of a potential energy gradient which is capable of inducing the movement of people? This question has been much debated. It would seem that we can speak of potential surfaces in the human realm but the creation of, say a potential surface of population in which the form of the surface is supposedly capable of giving rise to flows of people, requires some juggling with the population data. It is argued that, in terms of population, the influence one location has on another depends on the relative positions of the two and the distance between them; in any region, the potential influence of any one location depends on the total influence of all other locations. Thus, the population potential, V_i, of any one location may be defined as

$$V_i = \sum_{j=1}^{n} (P_j/d_{ij}).$$

It can be seen that the population potential is a function of the sum of the population of all other locations in the region, P_j, divided by the distance separating them from location i, d_{ij}. The computational process entails taking each location in the region and working out the ratio P_j/d_{ij}, then summing the values to give the population potential at i which will be in units of per person per unit distance. With population potential evaluated at each of the n locations in the region, a contour map can be constructed which depicts the form of the population potential surface. This basic concept of a potential surface in human geography can be adapted to other variables. Income of the population at a location can be used with distance to produce an income potential surface. Figure 4.20 shows the income potential surface for the USA in 1967 constructed from 3070 control points. Further modifications can be made. Instead of distance, transport costs between pairs of locations can be used with population or retail sales to produce a cost surface. Similarly, the problem of finding the optimum location for a factory given market locations, raw material sites, and transport costs—the Weberian problem of industrial location—can be solved by constructing a transport-cost surface and finding the lowest point on it.

Surfaces possess several features which may be used in their description. A contour joins lines of equal value on the surface and is technically

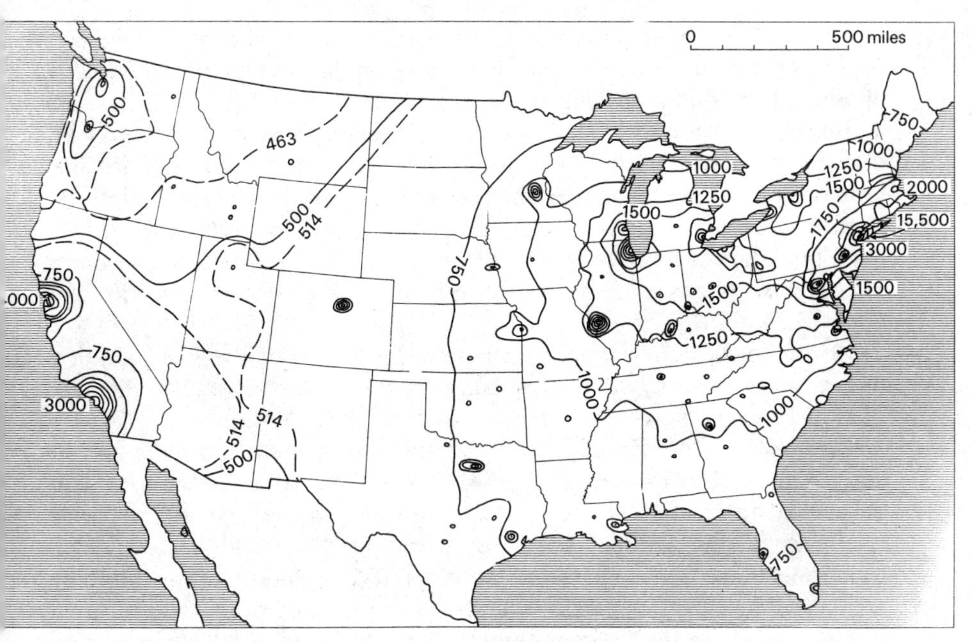

FIG. 4.20. The income potential surface for the USA in 1967. Units are millions of dollars per mile. Reprinted with permission of the author from W. Warntz (1973).

defined as the intersection of the surface of interest, say the ground, with a plane (horizontal surface). The intersection of several horizontal surfaces—usually with a regular vertical spacing between—with the ground gives rise to a series of contours. If the contours so formed are projected on to a plane surface, a topographical map is formed. A contour at a peak (in our example a mountain top but it could be say a local high value of population potential) is a point which is formed by a horizontal plane just touching the ground surface. A peak is a local maximum point, all surrounding points being at a lower elevation. A contour at a pit (depression) is also a point but in this case a local minimum one, all adjacent points being at a higher elevation. Peaks and pits are known as absolute extremum points. Other points, generally termed 'mixed extremum points', lie in topographical saddle points, adjacent points being higher in one direction but lower in another direction. Warntz (1973) distinguished passes—points that lie between two peaks, and pales—points that lie between two pits, but topologically these are identical. Through each point on a surface passes a line indicating the direction of a steepest gradient. Such slope lines lie at right angles to surface contours. Generally speaking a slope line runs up hill to a peak and down hill to a pit. Some slope lines join peaks—these are called ridge lines; some slope lines join pits—these are called course lines.

Surfaces may be divided into component areal units. Any surface can for example be divided into hills, each of which has its own peak and each of which has boundaries defined by course lines on which lie pits, passes, and pales. On the other hand, and quite independently, a surface may be divided into dales, each of which has its own pit and each of which has boundaries defined by course lines on which lie peaks, passes, and pales. Course lines and ridge lines divide a surface into territories, each territory's boundary usually consisting of two ridge lines and two course lines. All singular points extrema are found on the boundaries of territories; none lies within territories.

Movement of material across a surface, be it through a point, along a line, or over an area, will tend to run parallel, to converge, or to diverge. The vergency of flows on surfaces is shown in Table 4.2.

Much work has been directed to identifying land-surface units and to describing land form. Speight (1974) showed that land-surface form can be viewed in relation to at least two models. In the first model, basic land-surface units are seen as simply curved, mathematical surfaces which lack inflections; these units are considered in relation to surrounding elements—up slope, down slope, and either side—and slope changes are a prime criterion in their recognition. In the second model, the land surface is seen as a surface whose basic units show a repetitive or cyclical pattern across space; relief is fundamental to this model. The basic landscape components in the first model are called landform elements, and in the second model they are termed landscape patterns. Landscape elements go by various names including facets, sites, land elements, terrain components, facies, and land facets. Several measures have been used as descriptors of landscape elements; the most common are slope angle, slope curvature, and contour curvature. To put an element in a wider

Table 4.2
Vergency of slope lines

Class	*Feature*	*Vergency*
point	peak	divergent
	pit	convergent
	pass	mixed
	pale	mixed
line	course line	convergent
	ridge line	divergent
area	hill	divergent
	dale	convergent
	territory	mixed

Source: W. Warntz (1973).

locational perspective, slope length from a point on a profile upslope to the watershed, and the height difference associated with this distance are used. Speight (1974) derived several landform element descriptors from primary measures of altitude and slope. Workers have found that landscape elements seem to be arranged in definite ways to form landscape patterns which are variously named stows, recurrent landscape patterns, land systems, landform systems, relief units, and landscapes. Five main attributes are used to describe landscape pattern: landscape element composition; toposequences—the succession of elements found along a line of maximum slope; lineations, such as cuestas and joints, and networks, such as stream networks; planes of accordance; and relief and grain. The planes of accordance are two surfaces fitted to a landscape, an upper one just touching the major crests or summit surfaces, and a lower one just touching the major channels or valley floors. The distance separating the planes of accordance provided a typical measure of the relief of the landscape.

4.2.2. *Networks*

The movement of materials in the systems of both physical and human geography tends to become channelled into specific paths. The channelling of water flow in a drainage basin gives rise to streams, and the canalization of the flow of material, people, and information in an economic landscape leads to transport and communications routeways. Flow paths in a system usually combine to form a network which is susceptible of description and analysis. Networks in geographical systems are many and various; Haggett (1967) proposed a topological classification of them (Table 4.3.).

Planar graphs are real systems of lines on the earth's surface, whereas non-planar graphs contain paths which cross in plan but do not actually intersect, as in air routes which lie at different heights. Topology is the

Table 4.3
Topological classification of geographical networks

Planar graphs:		
Linear flow systems (movement along lines)	Paths	individual roads, rivers, stretches of coastline
	Trees	stream networks
	Circuits	transport and communication networks
Linear barriers (movement across lines)	Cells	county boundary network
Non-planar graphs		air route networks

Source: based on a table in P. Haggett (1967).

most abstract of geometries and deals with connections between points in terms of their relative, not absolute, location. Viewed topologically, networks are stripped down to their fundamental form and studied as topological graphs consisting of just two elements: vertices and edges. Vertices are also referred to as nodes, junctions, intersections, terminals, and zero-cells; edges are variously called links, sides, arcs, segments, branches, routes, and one-cells. Of the different kinds of geographical network, circuits and trees have been most studied and will be dealt with in a little more depth here; details of paths and cells can be found in Chorley and Haggett (1969) and Haggett *et al.* (1977).

Circuit networks. Road and railway networks, which form virtually the entire transport net in regional systems, contain closed loops or circuits; they thus are classified as circuit networks. Several topological measures have been devised to describe properties of circuit networks. The measures are based on three parameters: the number of edges in the network, e; the number of vertices in the network, v; and the number of subgraphs or subsidiary, unconnected graphs, s. From these parameters may be derived several measures of gross network characteristics. The simplest measure of network complexity is the Beta index, β, which is defined as the number of edges divided by the number of vertices

$$\beta = e/v.$$

In the example shown in Figure 4.21, which shows the south-east Manchester railway network in 1920 and 1975, the β index is 1.2 for 1920 and 0.875 for 1975. As is evident from these figures, the greater the β index the more complete the network. The cyclomatic number, μ, also

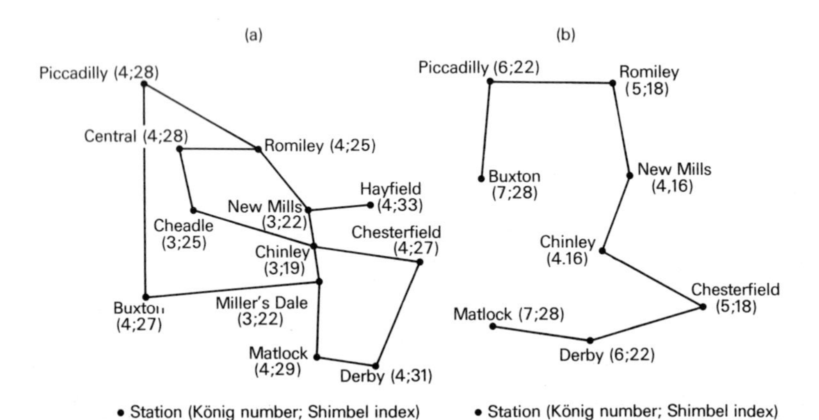

FIG. 4.21. The south-east Manchester railway network: (a) 1920; (b) 1975. Reprinted with permission, from M. G. Bradford and W. A. Kent (1977). *Human geography: theories and their applications*, Oxford University Press, Oxford.

measures network complexity

$$\mu = e - v + s.$$

In the Manchester railway example we have $\mu = 3$ for 1920 and $\mu = 0$ for 1975, and so the more complex the network the larger the cyclomatic number. The connectivity of a circuit network may be measured by the Alpha or redundancy index, α, which is defined as the ratio of the observed number of fundamental circuits in the network—this is given by the cyclomatic number—to the maximum number of circuits which is given by $2v - 5$

$$\alpha = (e - v + s)/(2v - 5).$$

In the south-east Manchester railway net $\alpha = 0.158$ for 1920 and $\alpha = 0$ for 1975. The larger the alpha value the more connected is the network. Another measure of network connectivity is the gamma index, γ, which is defined as

$$\gamma = e/\{v - (v - 1)/2\}.$$

The diameter of a network, *d*, is defined as the maximum number of edges in the shortest path between each point of vertices; it will be the lowest-value element in the shortest-path matrix of the network, that is, a matrix showing the number of links between pairs of vertices. The row totals in the shortest-path matrix are called Shimbel numbers and are a guide to the accessibility of vertices within the network, the lower the Shimbel number the more accessible the vertex. The largest element in each column of the shortest-path matrix is called a König number and indicates the centrality of a vertex within the network, the larger the König number the less central the vertex. The König number actually shows the maximum number of edges in the shortest path from any one vertex to any other vertex in the network. Table 4.4 shows the shortest-path matrix for the south-east Manchester railway network in 1975: the network diameter is 7; the most accessible vertices, with Shimbel numbers of 16, are New Mills and Chinley; New Mills and Chinley are also, with König numbers of 4, the most central vertices.

Tree networks. The land surface is characterized by a network of channels and streams. These networks do not contain circuits and are tree-like structures. Basic to the analysis of a stream network is the construction of a stream-ordering system which assumes a hierarchical relation among the branches of the tree-like net. The original ordering system was given by Horton (1945); Strahler (1952) modified it; and Melton (1959) showed that both Horton's and Stahler's scheme can be derived from elementary combinational analysis. A channel network

Table 4.4
Shortest-path matrix for the south-east Manchester railway (1975)

	Stations	1	2	3	4	5	6	7	8	*Row totals (Shimbel numbers)*
1	Buxton	—	1	2	3	4	5	6	7	28
2	Piccadilly	1	—	1	2	3	4	5	6	22
3	Romiley	2	1	—	1	2	3	4	5	18
4	New Mills	3	2	1	—	1	2	3	4	16
5	Chinley	4	3	2	1	—	1	2	3	16
6	Chesterfield	5	4	3	2	1	—	1	2	18
7	Derby	6	5	4	3	2	1	—	1	22
8	Matlock	7	6	5	4	3	2	1	—	28

Reprinted with permission, from M. G. Bradford and W. A. Kent (1977), *Human geography: theories and their application*, Oxford University Press, Oxford.

(Figure 4.22) represented by a set of straight-line segments is, topologically speaking, a finite-rooted tree: it has a root, a set of inner nodes or junctions at which three, and very rarely more, branches meet, a set of outer nodes or sources which give rise to one branch, and a node-linkage pattern which excludes the existence of closed loops within it. Such networks exhibit many regularities of structure. Let l be the number of links or branches, s the number of sources, j the number of junctions or forks, and n the total number of nodes. The number of links is the total number of nodes less one

$$l = n - 1.$$

Fingertip branches have one source; two branches lead to all other

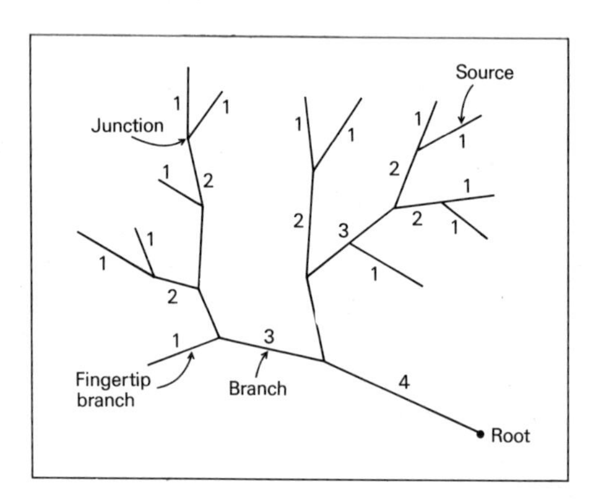

FIG. 4.22. A channel network.

successive sources: therefore

$$l = 2s - 1.$$

And it follows that

$$j + 1 = s.$$

If all the fingertip branches are pruned, then the links so removed are all first-order streams. A second pruning removes second-order streams; and so on. The number of prunings needed to cut the tree to just above the root determines the order of the network, at least according to this, the ordering scheme developed by Strahler. Other, not unrelated ordering schemes have been presented by Shreve (1966) and Scheidegger (1970). Hagget (1967) developed a version of network ordering applicable to road and other transport networks.

4.2.3. *Points*

A point pattern can be used to map any component of a geographical system which does not vary continuously over space. A point may represent individual people, farmsteads, towns, factories, plant species, certain landforms like drumlins, sink-holes, or mountain peaks, and so on. The description and analysis of settlements as a point pattern has had a long history in geography, the distinction between dispersed and nucleated rural settlement patterns receiving much attention (see Jones, 1964). Though a good portion of this research has been qualitative, mathematical and statistical techniques have long been employed to measure properties of point patterns and give information about the relations between points in geographical space. Thomas (1977) posed the questions a geographer might well raise about a point pattern: does the location of one point influence the location of other points? In other words, are the point locations dependent on one another? Or, more fundamentally, is there evidence in the pattern to suggest that the points are independent in space? In other words, do the points show a random distribution? The dependence or independence of point pattern is an important structural facet of a geographical system and may be analysed in several ways, the chief among which are quadrat analysis, nearest-neighbour analysis, and contiguity analysis.

A point pattern may be viewed in two ways: as a continuum or as a composite of several independent components. Quadrat analysis and nearest-neighbour analysis follow the continuum hypothesis, considering a random distribution of points as the centre of a spectrum which ranges from a clustered pattern to a regular pattern. Measures used as descriptors of the pattern's randomness include variance/mean ratio, the negative binomial k parameter, the redundancy value, and nearest-neighbour

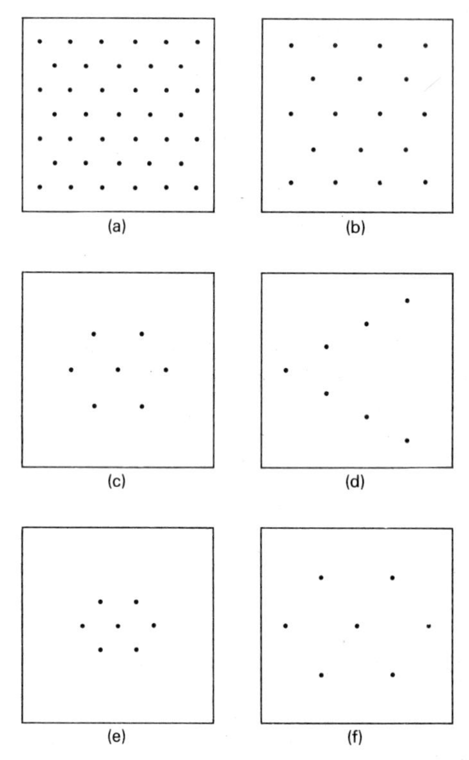

FIG. 4.23. Point patterns. Reprinted with permission of the author, from Dacey (1973).

statistic (see Thomas, 1977). Viewed as a combination of several independent components, a point pattern is described in terms of its pattern, density, and dispersion (Thomas, 1965). The pattern is the geometrical arrangement of points and is not influenced by the size of the area; the density is the frequency of occurrence of the points relative to the size of the area; these ideas are illustrated in Figure 4.23. A fashionable technique which reveals the arrangement of points is contiguity analysis.

4.3. Regional systems

4.3.1. *Regions as systems*

Regional systems consist of spatial structures embedded in flows of energy, matter, and information. The recognition of regional units is in fact a problematic process: the search for basic building blocks of man's spatial organization is one of the most persistent problems posed in geography. General categories of regional units were divided by Whittlesey (1954) on the basis of the number of criteria used in their delimitation: single-feature regions, multi-feature regions, and total regions.

Multi-feature regions, by far the commonest, are subdivided into formal regions, which are uniform throughout (homogeneous), and nodal regions, which are functional units in the sense that they are linked to a focal point by lines of communication. A good case can be made for viewing nodal regions as systems. A nodal region is a settlement—be it farm, village, town, or city—and the surrounding area to which it is tied by the organized movements of people, goods, finance, information, or by influence (Haggett, 1972, p. 247).

The advantages of regarding the nodal region as a basic regional-system unit have been put forward by Haggett (1972, p. 247–8): (1) the globe is becoming increasingly node-centred; (2) nodes are easily identified and mapped; (3) nodal regions facilitate the comparison of different parts of the world rather than their uniqueness and thus lay the foundations for general theories of human spatial organization; (4) nodal regions form a functional hierarchy of nested units ranging from a hamlet to the world level. Despite these persuasive arguments for its adoption as the basic unit of regional systems, the nodal region is susceptible of dispute. Chisholm (1975, p. 75) saw regions as a special case of the general problem of classification of data sets; he thus regarded them as abstractions which do not exist as real entities; this point is averred by Bunge (1962, p. 14–26) who believed that to seek regions as classes in geography is but a healthy, taxonomic phase common to all sciences. Though most geographers would be inclined to condone regionalization as classification, utterances of discontent would likely follow any revival of the suggestion that regions are real entities. But the notion of a nodal region as a functional unit implies that we are looking at a system of the real ilk rather than a mere abstract aggregate of classes as is the case with taxonomic units. The nodal region as a real thing may be unpalatable to many geographers because it brings to mind the organic analogy applied to cultures and civilizations by Danilevsky and Spengler, as well as the organic analogies once rife in many branches of geography (see Stoddart, 1967). What is looked over is that comparison with an organism can simply draw attention to the formal correspondence in the types of sub-systems found in organisms and in regional systems, as shown in Table 4.5. In this framework, the usefulness and existence of nodal regions as regional-system units gains felicity. Even so, of the sub-system units shown in Table 4.5, the system boundaries are most difficult to define. Indeed, the diffuse nature of regional-system boundaries creates the problem of closing the system to form a unit which can be subjected to analysis. A similar problem has been encountered in defining the boundaries of ecological units, whose boundaries are not easily defined in anything but an arbitrary manner for units at a higher level that individual organisms. But boundaries there are. Take a population. Each member of

Table 4.5
Formal correspondence of sub-system units in organism and regions

Sub-system type	*Organism*	*Region*
boundaries	unit membrane; skin cells	regional boundaries
transducers	mouth–anus; sensory cells	ports, airports; radar, satellites
processors	stomach; brain	industry; information-consuming population
storage units	fat deposits; memory	regional resources, libraries, and other information storage establishments

Based on an idea of Miller (1965).

a population carries genes and the total stored genetic material in the whole population is called the gene pool. Genetic material is exchanged within the population or deme. The size of such demes varies hugely from a very few individuals in a newly colonized area, through thousands in many animals, to millions in some small insects. Thus Pianka (1974, p. 66) has noted that, compared with cells and organisms, populations are abstract entities but are none the less real: the gene pool has continuity in time and space, for organisms in a particular population have a common immediate ancestry or are potentially able to interbreed or both. At the level of entire ecosystems it is not clear where system boundaries lie. Some advocates of the ecosystem approach see the ecosystem as a real thing. Rowe (1961) saw ecosystems as objects which lie above the level of organisms in the natural order of things. Similarly, Schultz (1969) stressed that entities such as ecosystems, which have well developed structural and functional characteristics, are autonomous and stand out as natural objects of study. Regional systems come in the same category.

Another way of showing that regional systems do have boundaries is to look more carefully at what exactly a boundary is. Platt (1969) has put forward several general properties of boundaries in nature, some of which are found in the boundaries of regional systems. Firstly, boundaries lie between regions of high interaction and regions of low interaction, where some variable such as interaction density is at a minimum; this principle is applied in the delimitation of geographical fields. Secondly, the boundary of one system property coincides with the boundaries of other system properties because boundaries are mutually reinforcing; in geographical terms, the degree of boundary coincidence is not as close as with biological systems but nevertheless there is often a degree of overlap, as in indicators of urban fields for instance (Green, 1955, is a case in point). Thirdly, boundaries usually contain gates, such as the mouth and ear, which allow the passage of matter, energy, and information. Analogies in regional systems include frontier posts, city gates, sea-ports, airports, and railway terminals.

4.3.2. *The hierarchy of regional-system units*

A hierarchical arrangement of nodal regions has been identified in many areas, one of the classic studies being made by Bracey (1962) in southern England. It has been questioned, however, whether, rather than forming structurally distinct systems, the hierarchical units are simply empirical classes superimposed for convenience on what in reality is a continuous functional relation between population size and the units in which people live (as measured by the number of functions they contain). Evidence for hierarchical breaks in an over-all continuous relation has been gathered, for instance, from Iowa (Berry *et al.*, 1962). In this study, settlements were grouped by factor analysis on the basis of number of functions, kinds of functions, and numbers of functional units; three district classes emerged from the analysis, viz. cities, with more than 55 functions; towns, with from 28 to 50 functions; and villages with between 10 and 25 functions. Hamlets were not included in the analysis.

The city itself can be divided into hierarchical units. In the American city, Berry (1959) recognized three basic city components: shopping centres, ribbon development (including the traditional shopping street and others catering mainly for the demands of a road), and specialized areas catering for special consumer demands; these units show a hierarchical structure. Garner (1966), using regression and co-variance analysis, identified the following hierarchy in Chicago: neighbourhood centres, providing convenience goods for people living locally; community centres, providing infrequently demanded goods for several neighbourhoods; and regional centres, supplying specialized goods for residents of large portions of the urban area. On average these centres provide 56, 37, and 24 functions respectively and Garner (1967) noted the comparability of these figures with those of villages, towns, and cities is rural Iowa and suggested that intra-urban hierarchies are a logical extension of the hierarchies developed under lower population density conditions.

Units of agricultural systems also seem to possess hierarchical organization, with units ranging from a field, through the farm, to an agricultural region. Each of these system units are related to one another and to systems other than agricultural ones. Munton (1969) had made a case for looking at farm systems, at least in present-day Britain, as a basic agricultural system unit: the farm system is a business unit for which decisions are made; the regional agricultural system consists of farms and its characteristics are difficult to analyse without recourse to farm data; farms are planned as entities and often individual field requirements are subordinated to an over-all farm policy; except in special cases like hill farms, government policy can be treated as a constant external influence and much of the variation between farm systems can be attributed to endogenous system properties. A similar case has been made by Chapman (1974) who showed that research into agricultural problems, especially those in the Third World, tends to be at two scales: at a small scale, crop technologists and the like look to improve yields; at a national scale, problems of total crop production, total fertilizer production, and so forth are considered. The intervening scale, that is farm management and farm economies, though it has not passed unanalysed, suffers from a paucity of conceptual models where emphasis is on the farm and the farmer.

4.3.3. *The structure and function of regional systems*

The components of regional systems are multifarious and can be viewed at any level of the regional hierarchy. Most models of regional systems focus on the national, regional, urban, or intra-urban levels. Models of one nation, one region, one city, or one city zone are known as spatially aggregated models and lack locational features: models which consider the distribution of state variables such as population among a set of zones within a nation, region, city, or city sector are called spatially disaggregated models and bear locational facets.

Spatially aggregated regional models. Many economics-based, regional models, such as the Susquehanna River Basin model (Hamilton *et al.*, 1969), Forrester's (1969) model of urban dynamics, and Watt *et al.*'s (1974) model of US Society, are spatially aggregated or have at most very crude locational components. The Susquehanna river basin was represented by eight sub-regions for each of which demographic, economic, and water systems were defined by suitable state variables. Watt *et al.*'s (1974) model considered state variables at two regional levels—a composite urban area and the rest of the United States. In Forrester's (1969) study a single region was represented by three groups of components, each of which consisted of three classes giving a total of nine state variables to define the system. The first group was new, mature, and declining enterprises; the second group was premium, worker, and underemployed housing; the third group was managerial-professional, labour, and underemployed workers. Such models, though essentially non-spatial, are not without interest to geographers and will be considered in more detail later in the book.

Spatially disaggregated regional models. Wilson (1974) has developed a spatially detailed regional model which seems potentially of more direct use to geographers than the single region models. Emphasis in Wilson's model is not on just one region nor on individual people, houses, places of work, or whatever, but on aggregate measures of people, houses, or whatever in a set of spatial zones. Interest centres on two interacting systems which are seen to form the regional complex—the spatial-demographic system and the spatial-economic system. The spatial demographic system consists of sub-systems representing the major activities of population—residence, work-place, use of services, use of transport. The economic system, or system of organizations, is composed of sub-systems representing the spatial economy as perceived by the population—housing, jobs, services, transport facilities, and land use. The system components interact (Figure 4.24); each may be further defined and measured. People may be classed into age, sex, social class, income, and other categories and measured by population size and so on. Components of the economic system may be classified by the activities they carry out—the brewing industry, the baking industry for instance, and measured by such things as jobs, products, and land use. We shall consider some components of this model in §7.3 under the guise of spatial interaction models; for a full and detailed exposition the reader should consult Wilson's (1974) work.

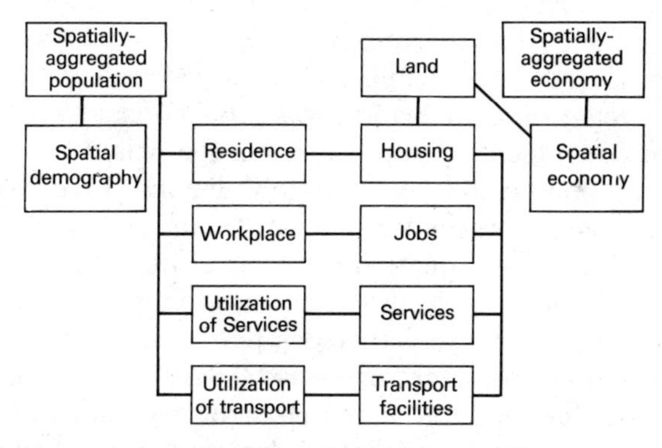

FIG. 4.24. Outline of the structure of a general regional system. Reproduced with permission, from *Urban and regional models in geography and planning* by A. G. Wilson. Copyright © 1974, by John Wiley & Sons Limited.

5 The Parsing Phase

> In natural science, I have understood, there is nothing petty to the mind that has a large vision of relations.
>
> ELIOT, *Mill on the Floss*

THE parsing phase of systems analysis involves the defining of relations between system components. Relations may be stated in many ways ranging from the verbal to the mathematical. At a verbal level we could say 'water enters the soil'; the same idea could be expressed as an equation describing the process of infiltration. The chief concern in this book is with relations as mathematical statements and, for the purpose of discussion, these will be divided into deterministic and stochastic types. A deterministic relation is a mathematical expression containing variables, parameters, and constants: an equation which stated that the annual increase of a population was proportional to population size would be deterministic. A stochastic relation is a mathematical expression containing variables, parameters, and constants, *and* a random component which describes a process on a probability basis: an equation which stated that the annual increase of a population was proportional to population plus some random component of population change would be stochastic.

Statistical relations could be separately defined. Like stochastic relations, statistical ones have random elements, but random elements arising from measurement error, equation error, or to the inherent variability of the variables and not to the inclusion of a random process. So the relation between population density, y (persons per hectare) and distance from Norwich city centre, x (kilometres)—$y = 60.92 - 9.93x$—is established by a statistical procedure and has an inbuilt error component. None the less, the relation defines exactly population density for any distance from the city centre and as such may be used deterministically, providing its limitations are borne in mind. Indeed, the calibration of many system models requires the determining of statistical relations between some of the system components and these are used in a deterministic fashion. For this reason, we shall consider statistical relations, geographical examples of which proliferated during the so-called quantitative revolution, as empirically based deterministic relations, and not as a variant of stochastic relations nor as relations in their own right.

It would be wrong to surmise that deterministic and stochastic relations form two exclusive classes: the division, though not illogical, is really an expedient for discussion: many stochastic processes can be couched in deterministic terms and vice versa.

5.1. Deterministic relations: the empirical base

Two distinct statements of deterministic relations may be isolated: the one expresses relations as cause-and-effect linkages between pairs of system components; the other expresses relations as input-output links between system components and leads to the formulation of system dynamics as a set of simultaneous, difference equations. The first statement is largely based on empirical work; the latter is largely founded on theoretical considerations and will be discussed in the next section. Empirically based relations include relations in which one variable is expressed as some function of another, the nature of the function being appropriate for the data to which it is fitted; they also include relations expressed as a degree of association between pairs of variables which can be used to build up the causal structure of a system, that is, the groups of most closely related variables.

5.1.1. *Functional relations*

A function is a numerical dependence of a quantity on one or more other variable quantities. The population of a city will vary in relation to city area. If city population is called y and city area is called x, then y is said to be a function of x if, for every value of x, the value of y can be calculated or observed. A function is written for the general case as

$$y = f(x).$$

A host of symbols are sometimes used in place of f, including F, ϕ, y, and g. The variable x, city area in the example, is known as the independent variable or argument. Because y, city population in the example, depends on the value of x it is called the dependent variable. A function can be portrayed graphically as a curve. Thus, supposing that city population varied as the square of city area, we should have

$$y = x^2$$

and the curve showing this relation is given in Figure 5.1. The relation between dependent and independent variables is usually obtained by statistically fitting a curve to observed pairs of values of dependent and independent variables. The procedures by which this is achieved are beyond the scope of this little book; interested readers should consult standard works such as Krumbein and Graybill (1965) and Taylor (1977). The relevant points of the procedure here are that pairs of observations can be drawn up as a scatter diagram and a regression line between x and y values placed in the best-fit position through the points. With x as the

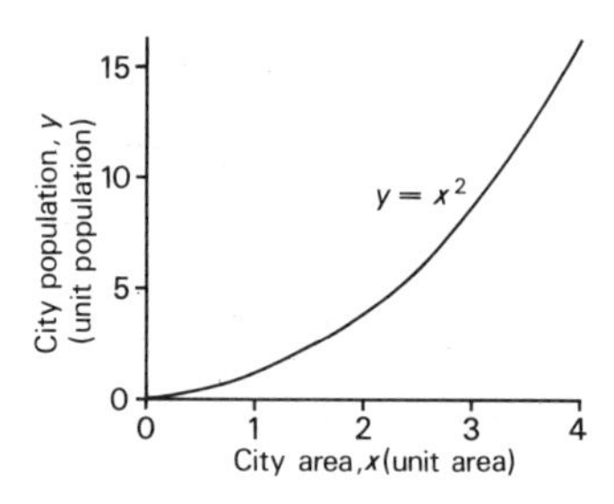

FIG. 5.1. Curve of the form $y = x^2$.

independent variable, it is assumed that x always produces a change in y, but y will not necessarily produce a change in x. In reality, empirically determined relations can be interpreted in several ways (Chorley and Kennedy, 1971, p. 24): x is the cause of y, which conclusion can only be based on a sound understanding of the processes involved; that x and y change in harmony but without any clear cause-and-effect connotations—this is autocorrelation and may arise if x and y are related through a third variable not included in the analysis; that the relation between x and y has arisen by chance, perhaps owing to some malpractice in the collection of data.

Several types of function may be fitted to empirical data. Unless there be any good reason for not doing so, a simple linear relation is usually assumed. This is a best-fit line relationship between x and y which takes the general form

$$y = a \pm bx$$

where the parameter a is the value of y when x is equal to zero, and the parameter b is the slope of the line, It has been found, for instance, that urban-land provision (the amount of urban land per person, the reciprocal or urban-population density) in 208 towns in England and Wales is directly and linearly related to social class, as measured by the percentage of occupied and retired males in classes I and II of the 1961 census, the relation taking the form

$$y = 7.38 + 0.60x$$

where y is urban-land provision (ha/1000 people) and x is social class (per cent therein) (Figure 5.2). Notice that the a parameter, 7.38, is the urban-land provision with no people in social classes I and II and the parameter b, 0.60, shows that for every unit increase in per cent in social classes I and II, urban-land provision increases by 0.6 units.

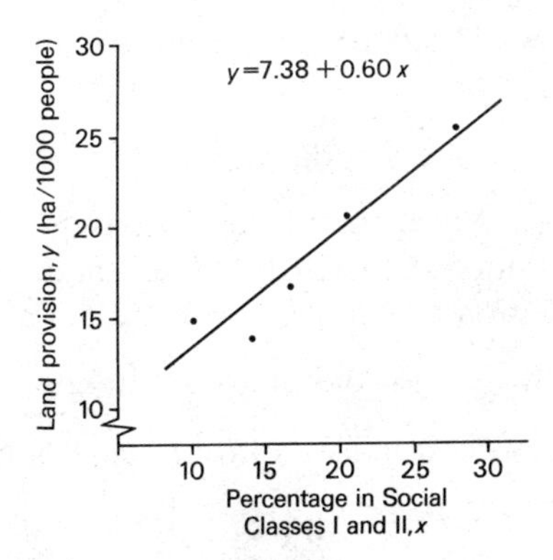

FIG. 5.2. The linear relation between urban land provision and social class for 208 towns in England and Wales. Reprinted with permission, from A. G. Champion (1972), 'Urban densities in England and Wales', *Area*, **4 (3).** 187–92.

Other functions can be fitted to data if they seem apposite. In geography, the following functions are commonly employed. A semi-logarithmic function takes the general form

$$y = a \pm b \log x,$$

in other words, the values of the dependent variable are a linear function of the common logarithm of the independent variable. In the case of urban-land provision in England and Wales, Champion (1972) found that urban-land provision (ha/person) was inversely related to the logarithm of town population (in thousands) (Figure 5.3.):

$$y = 30.19 - 7.64 \log x.$$

Very widely used are double-log functions in which the logarithm of the

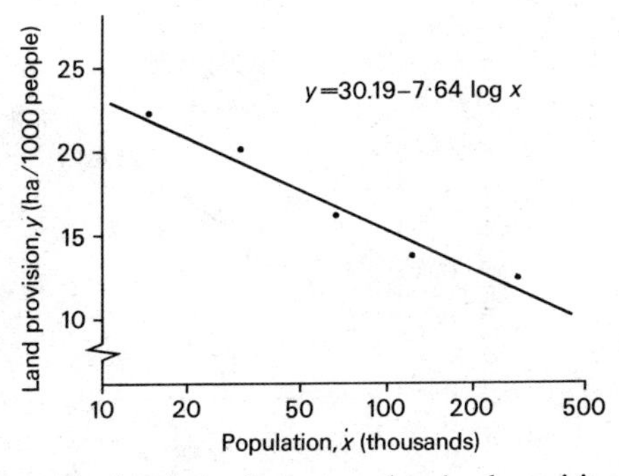

FIG. 5.3. The semi-logarithmic relation between urban land provision and town population for 208 towns in England and Wales. Reprinted with permission, from A. G. Champion (1972), 'Urban densities in England and Wales: the significance of three factors', *Area*, **4 (3).** 187–92.

dependent variable is a function of the logarithm of the independent variable

$$\log y = \log a \pm b \log x.$$

For instance, Thomas (1975) found the following relation between the central area of retail floorspace in British cities (millions of square metres) y, and the city population (millions), x (Figure 5.4):

$$\log y = \log 0.609 + 0.6711 \log x.$$

This type of equation is usually expressed as a power function, in the general case

$$y = ax^{b}$$

and, for the city central area retail floorspace example

$$y = 0.609x^{0.6711}.$$

Exponential functions are not rare in geography; they take the general form

$$y = ae^{\pm bx}$$

where e is the base of the Naperian logarithms. In general, population density in a city, y, declines exponentially with distance from the city centre, x, and may be described by the relation

$$y = ae^{-bx}$$

where a is the population denisty at the city centre and b is a parameter which ranges between 0.2 and 0.5 in American cities (Figure 5.5.). Sometimes a case can be made for fitting a quadratic function which has

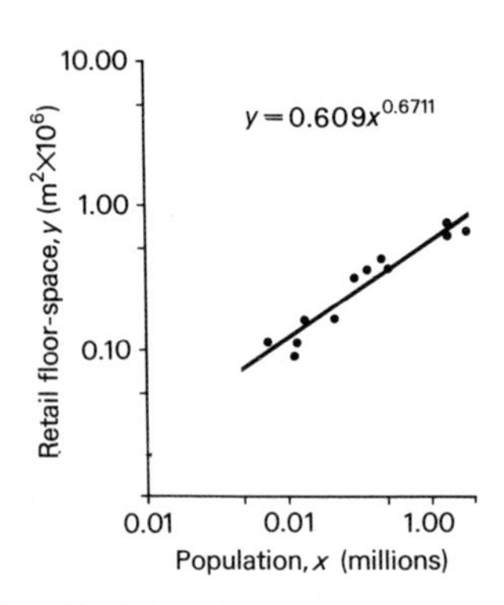

FIG. 5.4. The power function (double log) relation between the central area retail floorspace in British cities and city population. Based on a diagram in R. W. Thomas (1975).

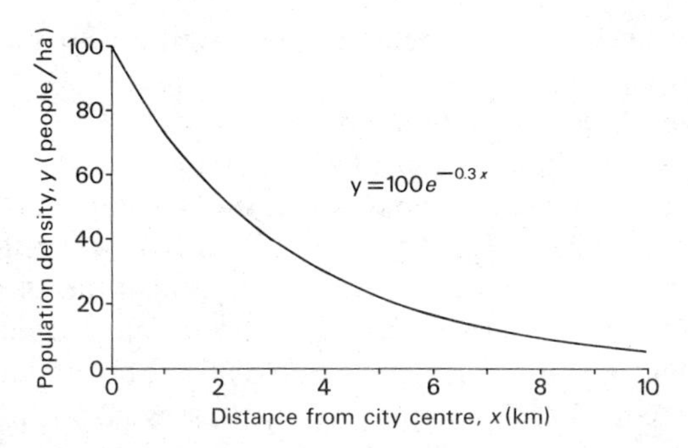

FIG. 5.5. An exponential decay curve showing population-density decline away from a city centre.

the form

$$y = a \pm bx \pm cx^2$$

and describes a parabolic curve. Koreleski (1975), working on loess soils near Kraków, Poland, found that the thickness of A horizons, y (cm), is related to slope angle, x (degrees) in the following way (Figure 5.6)

$$y = 63.0 - 9.84x + 0.54x^2.$$

A more detailed account of functions used in geography is given in Wilson and Kirkby (1975).

Correlation. It is sometimes useful to have a measure of the degree to which the independent variable, x, explains the dependent variable, y. In other words, to know how good x is as a predictor of y. One such measure is a parameter known as the coefficient of correlation, r. Values of r range from +1.0, for a perfect, direct correlation (y increases as x

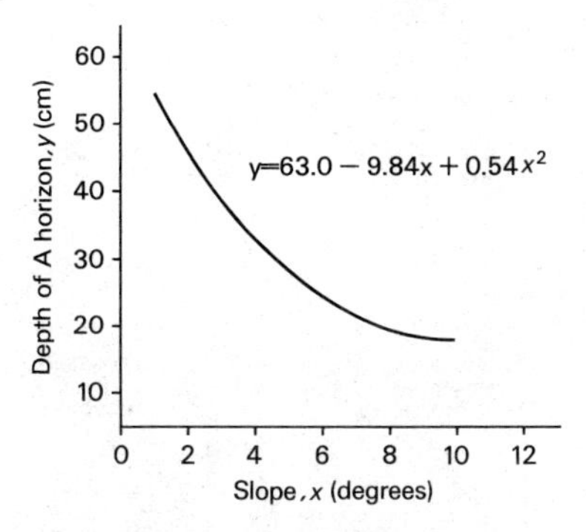

FIG. 5.6. The quadratic relation between the thickness of horizons and slope angle in loess soils near Kraków, Poland. Reprinted with permission, from K. Koreleski (1975), 'Types of soil degradation on loess near Kraków', *Journal of Soil Science*, **26 (1).** 44–52.

increases); through 0 for no correlation between x and y; to -1.0 for a perfect, inverse correlation (y decreases as x increases) (Figure 5.7). In statistical jargon, the correlation coefficient expresses the ratio of the explained variance of the data points with reference to the regression line, to the total variance; it is, as a percentage, the percentage of the y values' variance that is accounted for by a related variable x. In the example of city central area retail floorspace versus city population, the coefficient of correlation is +0.959 which suggests a strong, direct correlation between the two variables. Another measure, the coefficient of determination, the square of the correlation coefficient, expresses as a percentage the amounts of variance in y explained by the observed variance in x; in the city floorspace-size example $r^2 = 91.96$ per cent showing that a large percentage of the variance in the dependent variable is explained by the observed variance in the independent variable.

Owing to the fact that empirical relations and correlations are almost invariably based on sampled data, the relations and correlations between

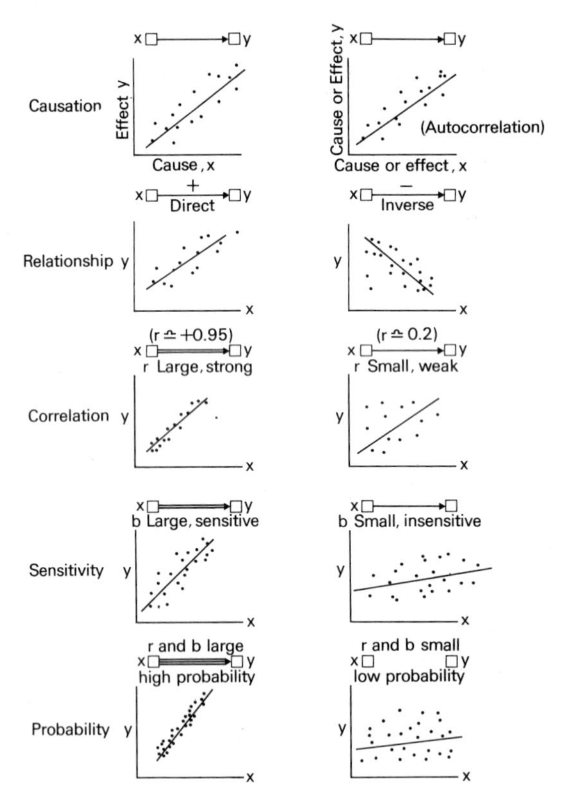

FIG. 5.7. Facets of correlation. Reprinted with permission, from R. J. Chorley and B. A. Kennedy (1971).

pairs of variables could arise by chance. The likelihood of their doing so is expressed as a probability, p. Thus $p=0.05$ means that there is a 5 per cent possibility that an association between two variables has arisen by chance; this would be known as the 95 per cent confidence or significance level because in 95 cases out of 100, the association can be expected to arise from factors other than chance. It is also common in the literature to find $p=0.1$, $p=0.01$, and $p=0.001$, the 90 per cent, 99 per cent, and 99.9 per cent significance levels respectively.

5.1.2. *Causal structures*

In a system consisting of more than two state variables, simple measures of relation between the variables can be put in a number of different ways. One way is to use multiple-regression equations and related multiple correlation coefficients; this topic is well covered in many geography books and will not be pursued here. Alternatively, correlation coefficients can be found between pairs of variables and a correlation system formed. Such correlation systems can be variously used. At a simple level, the signs of correlations can be employed to identify positive and negative feedback loops within the system; this shall be explored later in the chapter. A little more sophisticated than this is the expression of the strength of pairs of bonds between system components. At an even higher level of sophistication, the correlation system can be taken as the starting point in evaluating a system's causal structure.

One method of establishing groups of strongly related system components is linkage analysis. Given a square, symmetrical matrix of correlation coefficients between several variables (system components) linkage analysis, as developed by McQuitty (1957), can be performed to produce a grouping of these variables, the groups being known as typal structures. Typal structures are defined such that each member of a type is more akin to other members of that type than it is to a member of any other type. Yeates (1974, pp. 93–7) gives an example of linkage analysis applied to a matrix of correlation coefficients between twenty-one different socio-economic-cum-geographical variables measured for each of sixty-seven counties in the state of Florida in 1961 (Table 5.1). Following the McQuitty algorithm, the analysis runs as follows. The highest positive correlation coefficient in each column of the matrix is identified and marked. The highest entry in the matrix is then sought; in the Florida example this happens to be 0.99 and indicates a reciprocal relation between total population and total civilian population income: these two variables are thus the first two members of the first typal structure. The next step involves scanning the total population row and the total civilian population-income row and picking out those variables that are highest correlated to the variables in the typal structure. Population density and

Table 5.1
The correlation coefficient matrix for the Florida study

	TP	*TCPI*	*LA*	*TRM*	*AVF*	*TVFPS*	*DN*	*PCI*	% *under* $3,000	% *PI*	% 65+
Total population	1.00	***0.99***	0.44	0.35	***0.69***	0.46	−0.33	0.49	−0.48	0.43	0.03
Total civilian production income		1.00	0.46	0.32	0.65	0.47	−0.32	0.49	−0.46	0.38	−0.01
Land area			1.00	***0.61***	0.32	0.53	−0.43	0.46	−0.37	0.27	−0.05
Road mileage				1.00	0.25	***0.62***	0.04	0.29	−0.27	0.12	−0.07
Av. value per acre farmland					1.00	0.48	−0.58	0.51	−0.47	0.63	0.19
Total value farm prod. sold						1.00	−0.34	0.46	−0.33	0.22	0.08
Degrees north							1.00	−0.60	0.47	−0.49	−0.23
Per capita income								1.00	−0.77	0.57	−0.07
% under $3000									1.00	−0.72	0.14
% population increase										1.00	0.13
% 65 and over											1.00
% urban											
% rural farm											
Physical location											
% manufacturing											
% agriculture											
% mining and fishing											
% serv., prof., govt.											
Population density											
Road mileage per sq. mile											
% nonwhite											

(*continued*)

	% *U*	% *RF*	*PL*	% *M*	% *A*	% *MF*	% *T*	*PD*	*RMD*	% *N*
Total population	0.62	−0.31	−0.22	0.02	−0.36	−0.13	−0.06	***0.68***	0.10	−0.18
Total civilian production income	0.58	−0.26	−0.20	0.02	−0.32	−0.12	−0.07	0.60	0.03	−0.16
Land area	0.40	−0.28	−0.11	0.01	−0.12	−0.10	−0.18	0.00	−0.45	−0.13
Road mileage	0.35	−0.22	0.16	0.06	−0.27	−0.07	0.00	0.21	0.25	−0.03
Av. value per acre farmland	0.61	−0.35	−0.22	−0.12	−0.26	0.00	−0.04	0.62	0.13	−0.27
Total value farm prod. sold	0.41	−0.22	0.07	−0.05	0.06	−0.05	−0.36	0.21	0.01	−0.09
Degrees north	−0.44	0.37	***0.46***	***0.34***	−0.08	−0.04	***0.30***	−0.20	0.35	0.25
Per capita income	0.62	−0.48	−0.37	0.03	−0.14	−0.15	−0.20	0.35	−0.10	−0.18
% under $3000	−0.73	***0.66***	0.34	−0.16	0.44	***0.17***	−0.02	−0.36	0.06	***0.33***
% population increase	***0.66***	−0.53	−0.33	−0.04	−0.41	−0.08	0.08	0.37	−0.06	−0.37
% 65 and over	0.15	−0.09	−0.21	−0.34	0.10	0.09	−0.32	0.23	0.08	−0.26
% urban	1.00	−0.59	−0.25	0.02	−0.47	−0.11	−0.05	***0.54***	0.06	−0.26
% rural farm		1.00	0.34	−0.17	***0.63***	−0.16	−0.05	−0.27	0.01	0.16
Physical location			1.00	−0.08	0.24	−0.18	0.24	−0.19	0.22	0.19
% manufacturing				1.00	−0.35	−0.18	−0.15	0.02	−0.02	0.04
% agriculture					1.00	−0.20	−0.42	−0.33	−0.21	0.20
% mining and fishing						1.00	0.08	−0.13	0.05	−0.04
% serv., prof., govt.							1.00	−0.03	0.26	−0.06
Population density								1.00	***0.56***	−0.19
Road mileage per sq. mile									1.00	0.05
% nonwhite										1.00

Note. The highest positive correlation coefficient in each column is shown in bold italics.
Reprinted with permission of the author from M. Yeates (1974).

the average value per acre of farmland are most highly correlated with total population; but no variable other than total production is highest correlated with total civilian-population income. All these relations are called first relations. The next step is for all first relations to select second relations, if there are any, by the same procedure as in the previous step. In the Florida example, average value per acre of farmland is not highest correlated with any other variable whereas population density has highest correlated with it road-mileage density and percentage of population over sixty-five. The same process is then repeated to find third- and fourth-order relations but there are none in this first typal structure. A second reciprocal relation is then sought from variables other than those classified in the first typal structure; the procedure for assigning other variables to the second typal structure then runs as before. The entire process is repeated until all the variables form part of a typal structure. In Yeates's Florida Study, four typal structures were revealed (Figure 5.8):

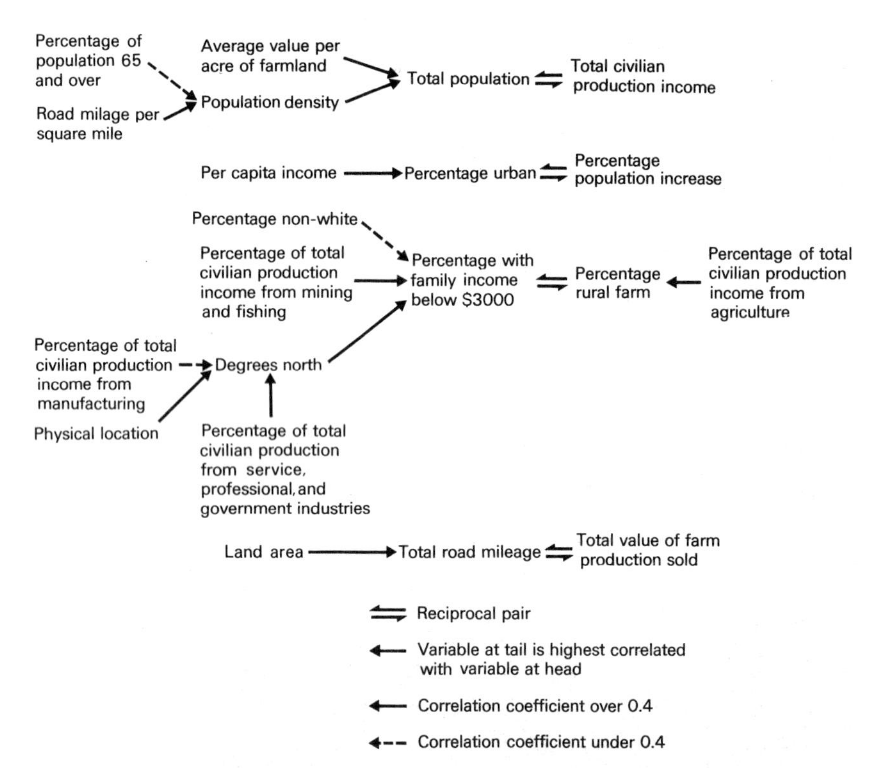

FIG. 5.8. Typal structures in the Florida socio-economic system of 1961. Reprinted with permission of the author and McGraw-Hill book Company, from M. Yeates (1974), *An introduction to quantitative analysis in human geography*, Figure 4–13, p. 97. Copyright 1974 McGraw-Hill Inc.

the first reflects the impact of the number of people—counties with large populations have high civilian production incomes and generally higher population densities which leads to a high road density; the second reflects the influence of urban affluence and growth, the most urbanized counties having the highest population growth rate and the highest per capita income; the third concerns the counties with rural deficiencies, including their location within the State; and the last involves factors pertaining to the sheer physical size of the counties.

Another method of establishing causal links within a system is due to Blalock (1964) and Simon (1954). The technique, called 'causal analysis', is to set up several causal models of a system and then to eliminate those which are inadequate in that they predict relations which are inconsistent with observed relations. In this wise, the causal structure of the system can be revealed. A causal model consists of the hypothesized direction of causal inference between state variables; observed relations are coefficients of correlation between pairs of state variables. Take the example, given by Hamilton *et al.* (1969), of four variables which form a model of regional relative wage change in the primary-metals industry for the Susquehanna river region in the USA. The four state variables in the study are: x_1, change in average wage by state for production workers from 1958 to 1963 in the primary metals industry; x_2, ratio of non-agricultural employment to urban plus rural non-farm population by State, 1960; x_3, fraction of State population that is rural farm, 1960; and x_4, average wage for all production workers in the state, 1958. Figure 5.9a shows the observed system relations as measured by correlation coefficients. The first causal model was set up as shown in Figure 5.9b.

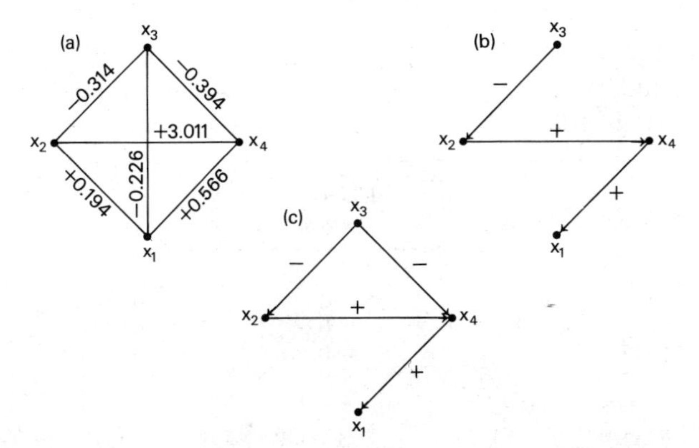

FIG. 5.9. Causal models of primary metal industry in the Susquehanna river region. Reprinted from *Systems simulation for regional analysis* by H. R. Hamilton *et al.* (1969), by permission of the MIT Press, Cambridge, Massachusetts. Copyright © 1969 by the Massachusetts Institute of Technology.

The arrows indicate hypothesized influences. It was predicted that the variable x_3, per cent rural, causes a decrease in variable x_2, urban employment ratio, which in turn causes an increase in x_4, average manufacturing wage in the state, which causes an increase in variable x_1, the wage change. Although no direct link between the per cent rural and wage change exists, a positive, indirect link through variables x_2 and x_4 is hypothesized. The next step is to use the observed correlation coefficients to make a set of predictions implied by the first causal model. Specifically, the values are predicted of correlation coefficients between each pair of variables which could be related but are not. This means r_{34}, r_{31}, and r_{21} must be predicted as these correspond to the unconnected pairs of variables x_3 and x_4, x_3 and x_1, and x_2 and x_1. For example, the model shows no direct link between pair x_3 and x_4 but some relation is to be anticipated owing to the indirect effect of x_3 on x_4 through the intervening variable x_2. The expected amount of correlation between x_3 and x_4 is simply equal to the product of the correlation coefficient between variables x_3 and x_2, that is r_{32}, and the correlation coefficient between variables x_2 and x_4, that is r_{24}. In other words the model predicts a correlation coefficient of $r_{34} = r_{32}r_{24}$. Substituting the appropriate values,

$$r_{34} = (-0.314)(0.311) = -0.098.$$

To see how good a prediction this is, −0.098 is compared with the observed value of r_{34} which is −0.394. So the actual relation between variable x_3 and x_4 is significantly more negative than the model would suggest. This process is repeated for other pairs of variables and, as can be seen in Table 5.2, the results are not good. The next step is thus to try to improve on the first causal model. The findings shown in Table 5.2.

Table 5.2
Two models of regional differences in wage changes in primary-metals industries

Predictions	*Degrees of fit expected*	*Actual*	*Difference*
Model 1.			
$r_{34} = r_{32}r_{24} = (-0.314)(0.311) =$	−0.098	−0.394	0.296
$r_{31} = r_{32}r_{24}r_{41} = (-0.314)(0.311)(0.566)$	−0.055	−0.226	0.171
$r_{21} = r_{24}r_{41} = (0.311)(0.566)$	0.176	0.194	0.018
Model 2			
$r_{31} = r_{34}r_{41} = (-0.394)(0.566) =$	−0.223	−0.226	0.003
$r_{21} = r_{24}r_{41} = (0.311)(0.566) =$	0.176	0.194	0.018

Reprinted from *Systems simulations for regional analysis* by H. R. Hamilton *et al.* (1969) by permission of The MIT Press, Cambridge, Massachussetts.

show that the inclusion of a direct link from per cent rural to average manufacturing wage improve the results. The second causal model can hence be assumed to fit adequately the real-world situation. However, and this is one flaw with this type of analysis, the second model is not necessarily the only one which would give results consistent with the data.

5.2. Deterministic relations: the theoretical base

5.2.1. *General formulation*

The general theoretical relations between system components can be derived from a formal definition of a system and its properties. A system is associated with, or discriminated by, a set of attributes—let us call them v_1, $v_2 \ldots v_m$—and a set of relations between these attributes (Figure 5.10); the defining attributes or variables are called terminal variables and the defining relations are called terminal relations. A concrete example should help to clarify this bland statement. The herbivore tropic level in a grassland ecosystem can be thought of as a system. The terminal variables associated with the herbivore system would include the grazing of the herbivores on plants (herbivory), the respiration of the herbivores, the predation of the herbivores by carnivores (carnivory), and the mortality of the herbivores: four terminal variables in all, v_1 to v_4 (Figure 5.11). If terminal variables are separable into input variables ($z_1, z_2 \ldots z_p$) and output variables ($y_1, y_2 \ldots y_q$), as they are in the example, and as indeed they are in most geographical systems, the system is said to be directed or orientated. In the case of the grassland herbivores, herbivory is an input variable, $z_{\text{herbivory}}$, and respiration, carnivory, and mortality are all output variables, $y_{\text{respiration}}$, $y_{\text{carnivory}}$, and $y_{\text{mortality}}$ respectively (Figure 5.12). If the inputs and outputs can be related by a function, then the system is a functional system. If inputs and outputs vary with time, the system is a time system. In a time system, the output may be defined by a response or output function, which we shall call g, and depends on the system state and the inputs. The state of the system, which in general is defined by state variables x_1, $x_2 \ldots x_n$, might in our example be the standing crop of the herbivore level, $x_{\text{herbivores}}$ (measured as say kilocalories per square metre).

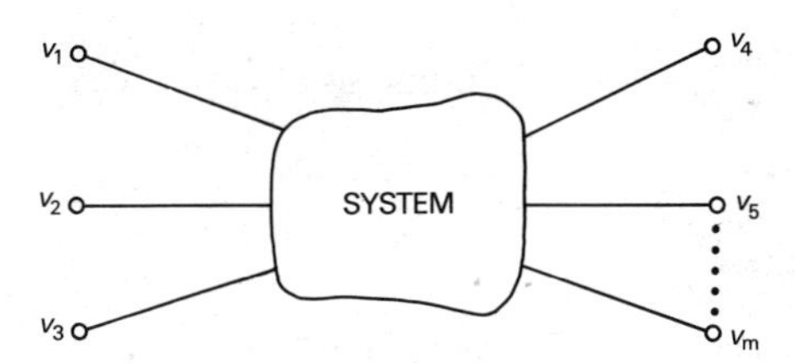

FIG. 5.10. A formal representation of a system. The vs are a set of system attributes.

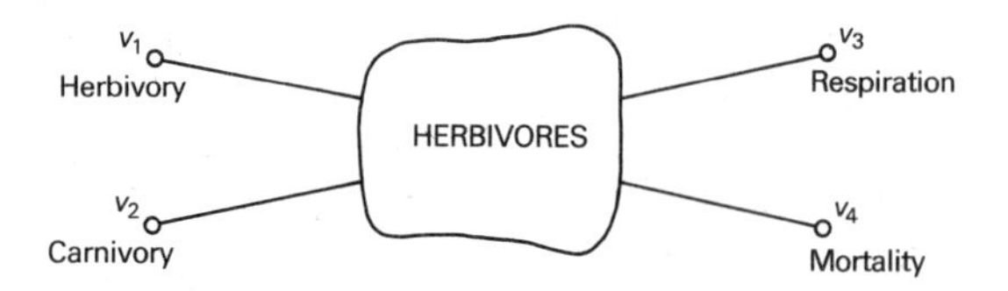

FIG. 5.11. A grassland ecosystem represented by four terminal variables.

So we have

$$\text{output} = \text{g(state, inputs)}. \tag{5.1}$$

Or, in the example, the respiration output would be

$$y_{\text{respiration}} = g(x_{\text{herbivores}}, z_{\text{herbivory}}).$$

A similar output equation could be written for mortality and carnivory

$$y_{\text{mortality}} = g(x_{\text{herbivores}}, z_{\text{herbivory}})$$

$$y_{\text{carnivory}} = g(x_{\text{herbivores}}, z_{\text{herbivory}}).$$

If the function, g, is known, an output function may be written for each output variable. For instance, if it is assumed that each of the three outputs is proportional, by factors a_1, a_2, and a_3 respectively, to the system state, $x_{\text{herbivores}}$, then the following output functions may be written

$$y_{\text{respiration}} = a_1 x_{\text{herbivores}}$$

$$y_{\text{carnivory}} = a_2 x_{\text{herbivores}}$$

$$y_{\text{mortality}} = a_3 x_{\text{herbivores}}.$$

The state of the system may change with time. As a general rule, given the state of a system at a particular time, the system state after an interval of time has elapsed, will be determined by: (1) the state in which the

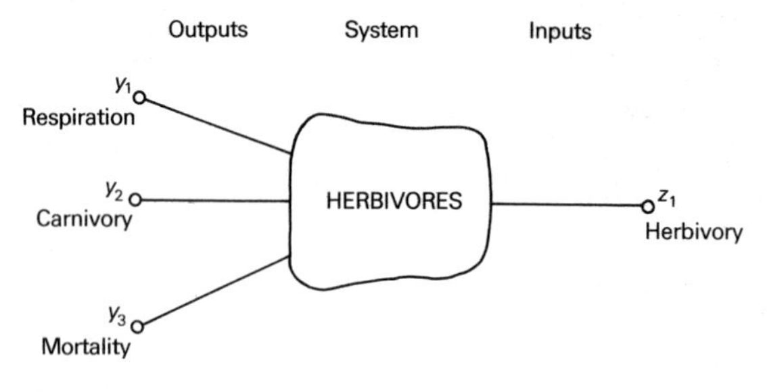

FIG. 5.12. A grassland ecosystem with terminal variables separated into input and output variables.

system started at the beginning of the time interval; (2) a function which defines the inputs and outputs during the time interval; and (3) the duration of the time interval. Using the following symbolism for the herbivore system

$$x_{\text{herbivores},\,t} = \text{state at start of time interval, time } t,$$

$$x_{\text{herbivores},\,t+1} = \text{state at end of time interval, time } t+1,$$

$$\Delta t = (t+1) - t = \text{the duration of the time interval,}$$

we may write

$$x_{\text{herbivores},\,t+1} = x_{\text{herbivores},\,t} + f(\text{outputs, inputs})\Delta t. \tag{5.2}$$

The outputs in the function f(outputs, inputs) depend, as determined by the output function (equation 5.1), on system state and inputs, so we may express equation 5.2 as

$$x_{\text{herbivores},\,t+1} = x_{\text{herbivores},\,t} + f(\text{state, inputs})\Delta t. \tag{5.3}$$

The function f(state, inputs) is called a state transition function; this is because it determines how the state changes from one time to another. A system for which both state transition and output functions exist is termed a dynamical system; most systems of interest to geographers come under this category. In the herbivore system the state transition function simply involves listing all input and output terms

$$f(\text{state, inputs}) = \underbrace{z_{\text{herbivory}}}_{\text{inputs}} \underbrace{- a_1 x_{\text{herbivores}} - a_2 x_{\text{herbivores}} - a_3 x_{\text{herbivores}}}_{\text{outputs (as determined by system state).}}$$

So we may now write equation 5.3 as

$$x_{\text{herbivores},\,t+1} = x_{\text{herbivores},\,t} + (z_{\text{herbivory}} - a_1 x_{\text{herbivores}} - a_2 x_{\text{herbivores}} - a_3 x_{\text{herbivores}})\Delta t. \tag{5.4}$$

Equations of this kind, which are called difference equations and describe the state dynamics of a system, are commonly encountered in the literature in a different guise which can be derived thus. Subtracting $x_{\text{herbivores},\,t}$ from both sides of equation 5.4 gives

$$x_{\text{herbivores},\,t+1} - x_{\text{herbivores},\,t} = (z_{\text{herbivory}} - a_1 x_{\text{herbivores}} - a_2 x_{\text{herbivores}} - a_3 x_{\text{herbivores}})\Delta t.$$

The term on the left-hand side, $x_{\text{herbivores},\,t+1} - x_{\text{herbivores},\,t}$, is the difference in the herbivore standing crop at the start of, and the end of, the time interval; it is thus the change in the standing crop during the time interval and may be more concisely written $\Delta x_{\text{herbivores}}$, the Greek capital delta, Δ, being used to denote a change. We now have

$$\Delta x_{\text{herbivores}} = (z_{\text{herbivores}} - a_1 x_{\text{herbivores}} - a_2 x_{\text{herbivores}} - a_3 x_{\text{herbivores}})\Delta t.$$

Dividing both sides of this equation by the time interval Δt yields

$$\frac{\Delta x_{\text{herbivores}}}{\Delta t} = z_{\text{herbivores}} - a_1 x_{\text{herbivores}} - a_2 x_{\text{herbivores}} - a_3 x_{\text{herbivores}}. \tag{5.5}$$

The term $\Delta x_{\text{herbivores}}/\Delta t$ is the change in the standing crop divided by the time interval and would be expressed in units such as kcal/m^2/yr; it thus represents the rate at which the system state changes and is known as the time rate of system change. It is in the style of equation 5.5, with the time rate of system change equated with the state transition function, that the information contained in equation 5.3 is commonly presented in books and papers dealing with systems analysis.

In actual fact, the grassland ecosystem might be seen to consist of three trophic levels—producers, herbivores, and decomposers. The standing crop of each trophic level is a state variable which may be designated x_1 (producers), x_2 (herbivores), and x_3 (decomposers). The state of the grassland ecosystem as a whole is uniquely defined at any time t by the values of the state variables at that time—$x_{1,t}$, $x_{2,t}$, and $x_{3,t}$. The three trophic levels are linked by inputs and outputs of energy (Figure 5.13). If state transition functions are drawn up for each state variable we end up with three equations which together describe the dynamics of the grassland ecosystem; they are

$$\begin{aligned} x_{1,t+1} &= x_{1,t} + (z - a_1 x_1 - a_2 x_1 - a_3 x_1)\Delta t \\ x_{2,t+1} &= x_{2,t} + (a_2 x_1 - b_1 x_2 - b_2 x_2)\Delta t \\ x_{3,t+1} &= x_{3,t} + (a_3 x_1 + b_2 x_2 - c_1 x_3)\Delta t. \end{aligned} \tag{5.6}$$

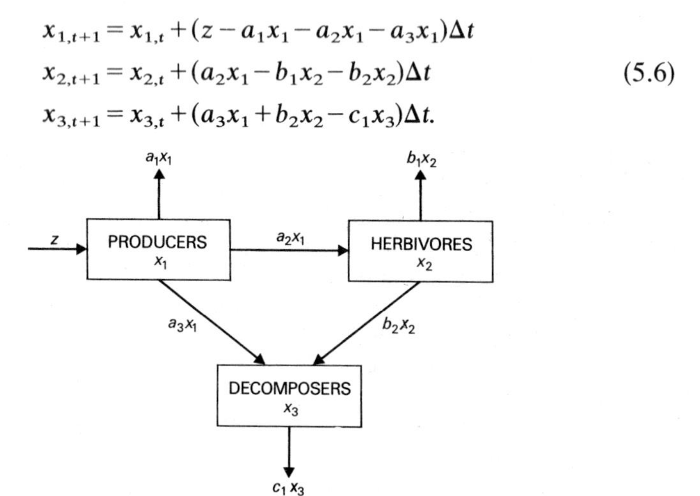

FIG. 5.13. Flows and storages of energy in a grassland ecosystem.

It can be seen that a change in any one state variable effects a change in other state variables; this demonstrates in mathematical terms that, through the interrelations between its components, a system acts as a single unit.

5.2.2. *Growth relations*

Growth relations are found in many fields of geography. The simplest case is the growth of a single state variable, x, which might represent the size of a human population. The dynamics of a single-component system may be described by the equation

$$x_{t+1} = x_t + f(x_t)\Delta t \tag{5.7}$$

which, if x is population, states that the population of the system at time $t+1$ depends on the population of the system at time t, a state transition function, $f(x_t)$, and the length of the time interval, Δt. The possibilities of this relation may be demonstrated by applying a device called a Taylor series expansion to the state transition function to produce

$$x_{t+1} = x_t + (ax_t + bx_t^2 + cx_t^2 + \ldots)\Delta t. \tag{5.8}$$

Retaining only the first term of the state transition function gives the growth equation

$$x_{t+1} = x_t + (ax_t)\Delta t \tag{5.9}$$

which indicates that the increase in the population between time t and time $t+1$ is proportional, by a constant a, to the size of the population at time t. If infinitesimally small time intervals are used, the equation can be solved to produce an expression which defines the population size at any time, x_t; the expression, in which time t is treated as a continuous variable that may adopt any value greater than or equal to zero, is

$$x_t = x_0 e^{at} \tag{5.10}$$

where x_0 is the initial population (that is, the population at $t = 0$) and e is the base of natural logarithms, 2.71. For positive values of the constant a, the population will grow exponentially; for negative values of a, the population will decline exponentially; for $a = 0$, the population will remain unchanged (Figure 5.14). This exponential growth law describes very well the individual growth of certain bacteria and animals; in demography it describes the unrestricted growth of plant and animal populations; in sociology it is the law of Malthus; it describes the growth of human and geographical knowledge; and has many other applications.

Retaining the first two terms of the state transition function in equation 5.8 and placing a minus sign in front of the second term gives a growth

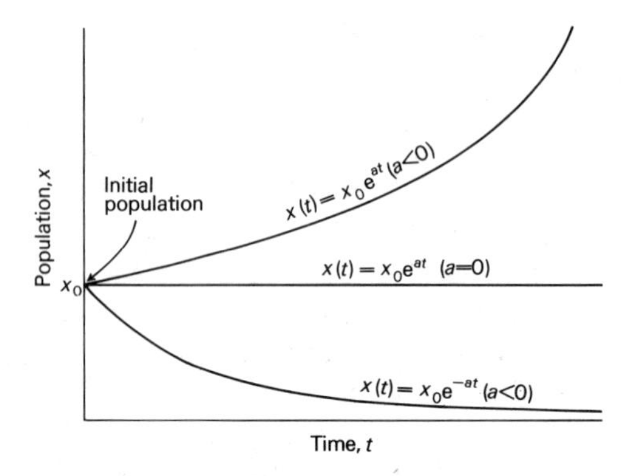

FIG. 5.14. Exponential growth and decline of a population.

equation with different properties to the first

$$x_{t+1}=x_t+(ax_t-bx_t^2)\Delta t \tag{5.11}$$

or, as it is commonly arranged

$$x_{t+1}=x_t+\left\{ax_t\left(1-\frac{b}{a}x_t\right)\right\}\Delta t. \tag{5.12}$$

Using infinitesimally small time intervals, equation 5.12 can be solved and the result is an expression which describes the logistic growth of the population (Figure 5.15): the bx_t^2 term counteracts the infinite, exponential increase of the ax_t term to produce an S-shaped or sigmoid curve in which the population attains an upper limiting value (a sort of system capacity) set by the value b/a; this is Verhulst's law which describes the growth of human populations with limited resources.

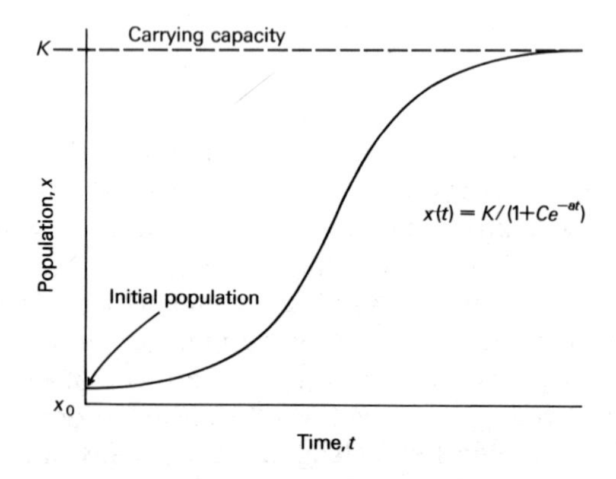

FIG. 5.15. A logistic curve of population growth. As the parameter b is set equal to a/K, it follows that $b/a=1/K$.

Growth relations for systems comprising two state variables are more complex. We have met the example of two interacting populations, the dynamics of which were traced on a phase plane (Figure 1.2). The interaction of urban and rural populations, and any other two-component system, may be studied in the same way. Growth relations in systems with large numbers of components are difficult to isolate; the effect of their over-all simultaneous action on system state can only be seen by simulating the behaviour of the system on a computer or locating mathematically stable and unstable nodes in the system's phase space.

The laws of allometry are another kind of growth relation. The law of allometric growth may be explained in the following manner. Let the variable x stand for the income of a region, and let the variable y stand for the income of the working class in that region. The absolute rate of increase of the regional income, which we shall set at £1000 per year, is written as $\Delta x/\Delta t$. The rate of relative change of the regional income, that is the increase calculated as a fraction of the total income, is written $\Delta x/\Delta t \cdot 1/x$. So, for a region whose income in 1975 was £50 000, the relative income increase would be $1000/50\,000 = 0.02$ per year or, as a percentage increase, 2 per cent per year. The absolute increase in income of the working class, which we shall set at £50 per year, is written $\Delta y/\Delta t$; and the relative increase in working class income is written $\Delta y/\Delta t \cdot 1/y$. So, if the income of the working class in 1975 was £5000, the relative increase would be $50/5\,000 = 0.01$ per year or 1 per cent per year. According to the allometric law, the relative increase in the income of the working class is proportional, by a constant b, to the relative increase of the regional income at all times: we may write

$$\frac{\Delta y}{\Delta t} \cdot \frac{1}{y} = b \frac{\Delta x}{\Delta t} \cdot \frac{1}{x} \tag{5.13}$$

relative increase in working class income	constant	relative increase in regional income.

In our example we have $0.01 = b \times 0.02$ so the constant b has the value 0.5; this means that the income of the working class will always increase at half the rate of the income for the region as a whole. We can perhaps gain a better picture of this if we write equation (5.13) as

$$\frac{\Delta y}{\Delta t} = b \frac{y}{x} \frac{\Delta x}{\Delta t} \tag{5.14}$$

which states that the working class, y, takes from the increase in the regional income, $\Delta x/\Delta t$, a share which is proportional to the relative income of the working class and the region, y/x; b, whose value in the example was 0.5, indicates the capacity of the working class to seize its share of income. Equation 5.13 is usually expressed as a power function in the form

$$y = ax^b$$

where the constant a is the income of the working class corresponding to a regional income of £1. In the case we have studied, this equation is similar to Pareto's Law which states that the number of people gaining a certain income, y, is proportional, by the parameter b, to the national income, x. In our example b would vary with different income groups: if b is less than one, the income of a particular income group increases less rapidly than the regional income increases; if b is one, the income group's income increases at the same rate as the regional income; and if b is greater than one, the income group increases faster than the regional income.

Many tentative proposals about allometric growth relations are to be found in Thompson's (1917) book *On Growth and Form* and the later synthesis of Huxley (1932) *The Problems of Relative Growth.* Recent work, such as that of Narroll and Bertalanffy (1956) and Gould (1966), has shown that the allometric law, and modifications of it, is applicable to wide-ranging problems which concern size relationships between parts of a system and between the parts and the whole system. Systems studied in this manner include drainage areas, animals, volcanoes, meanders, and urban areas. In urban geography, allometric principles first appear, albeit implicitly, in Zipf's (1949) derivation of the rank-size rule which explains city size-frequency distributions. Zipf's rule has subsequently been shown by Nordbeck (1971) to be equivalent to the allometric-growth law. Nordbeck (1965, 1971) successfully used the allometric law to study the relationship between urban areas and urban population in Sweden. Using similar principles, an interesting study of some functional characteristics of British city central areas was made by Thomas (1975). He found the following values of b for equations relating city population with, respectively, office floorspace, industrial floorspace, warehouse floorspace, and retail floorspace: 1.1975, 1.3097, 1.6001, 0.6711. He interpreted these results, assuming that size difference in space can be translated into time differences, as indicating that office, industrial, and warehouse floorspace tend to decentralize with time (they grow more rapidly than the population) whereas retail floorspace tends to decentralize (grows less rapidly than the population). In other words, within the population range studied, cities' central areas change from dominantly retailing areas when small to dominantly warehouse, office, and industrial areas when large. Properties of channels and drainage basins often exhibit allometric properties in that they are related by a double-log equation. Thus Leopold *et al.* (1964) found the relation

$$L = aW^{1.01}$$

between meander wavelength, L, and channel width W. That $b = 1.01$ is to be expected since both L and W are of dimension 1 (just length). On the other hand, with river length, L, plotted against river-drainage area A, which is of dimension 2 (length squared) we should expect b to be about 2, and indeed Nordbeck (1965) found the relation

$$A = 0.104L^{2.0009}.$$

5.2.3. *Feedback relations*

Dip into almost any book on systems analysis and the terms 'negative' and 'positive feedback' are bound to turn up. Positive feedback is brought about by what Maruyama (1960, 1963) called deviation-amplifying processes which, once system state starts to change, keep it changing. Exponential growth is an example of it. Negative feedback is brought about by what Maruyama (1960, 1963) called deviation-damping processes which, by counteracting the effect of any change in a system, keep the system state steady. Logistic population growth, once the carrying capacity has been reached, is an example of it.

Though the terms 'positive' and 'negative feedback' express specific relations between sets of system components and, if used prudently, can be useful bits of jargon, they can give in at least three ways an over-simplified or even false impression of the state-changing relations in a system. In the first place, certain recurrent types of relations between system components occur which, although they could be classed as essentially negative or positive feedback relations, would lose considerable information about their nature if they were so classed.

For instance, two types of relation can produce exponential growth: the first is the simple exponential model, wherein the time rate of change of the state variable is equal to the inputs minus the outputs, both of which are proportional to the state variable; the second is a pseudo-logistic growth model with a high energy source, wherein the time rate of change of the state variable is given by input and outputs, outputs being proportional to the state variable and a high energy input. And at least three distinct types of relation can produce logistic-type growth (see Odum, 1976).

Secondly, in any complex system—complex, that is, by virtue of the sheer number of state variables—it is likely that system changes will be rung by a host of system relations, some deviation-amplifying, some deviation-damping, acting in concert. Admittedly, the system as a whole may tend to show a stable state, which implies a preponderance of negative feedback relations over-all; or may be growing or declining, implying over-all positive feedback relations in the system. Figure 5.16 shows a system representation of an Arctic tundra ecosystem. The links between system components form a series of negative-feedback loops and the system is stable. Figure 5.17 shows a system representation of the land affected by earth-dam seepage in the Washita river basin, Oklahoma. The links between the system components form a series of positive-feedback links so that once seepage of water from an earth dam starts, the system keeps changing, presumably towards a new stable state.

Thirdly, there is a danger of mentally associating negative-feedback relations with a stable system and positive-feedback relations with a system moving away from stability. Certainly, it is erroneous to equate positive feedback with the idea of a vicious circle. A vicious circle is usually taken to mean a chain of vicious effects which spirals in a detrimental way. Although this is an instance of positive feedback,

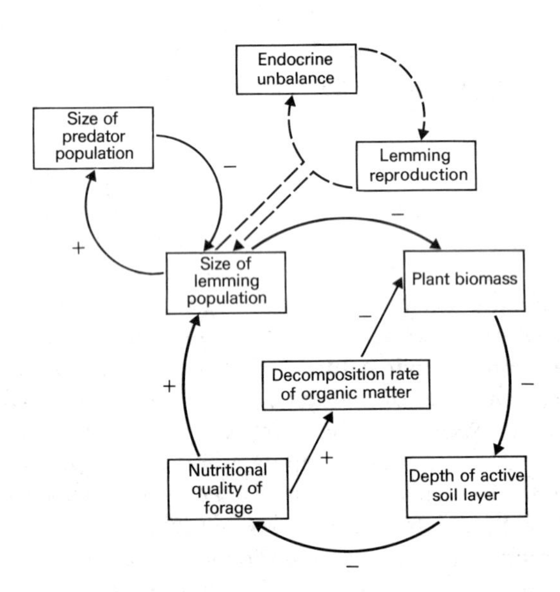

FIG. 5.16. System representation of an Arctic tundra ecosystem showing negative feedback loops. Reprinted with permission, from A. M. Schultz (1969), 'A study of an ecosystem: the arctic tundra', in *The ecosystem concept in natural resource management* (ed. G. M. Van Dyne). Academic Press, New York.

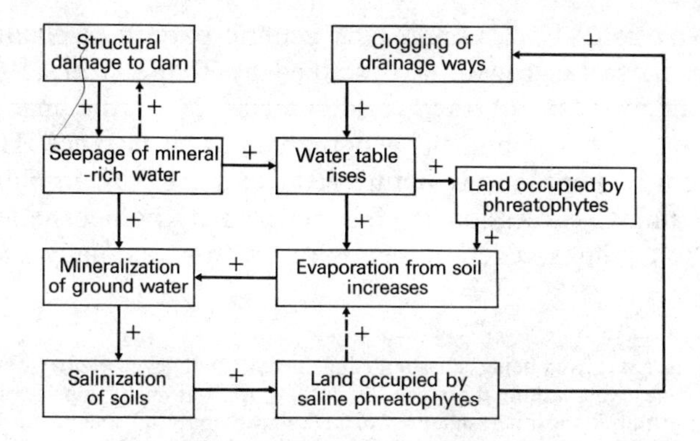

FIG. 5.17. The effects of earthdam seepage in part of the Washita river basin, Oklahoma. Based on C. Yost and J. W. Naney (1975).

positive feedback is also fundamental to all growth processes where it is responsible for rapid initial growth of systems and in this regard does not constitute a vicious circle.

The interplay of positive- and negative-feedback relations in a system can be subtle. Paradoxically, both types of relation can operate simultaneously to maintain the over-all stability of a system amidst perturbations in the environment. In the biological parlance, homeostasis is that group of system-stabilizing relations which are characterized by negative feedback; homeorhesis is that group of system-stabilizing relations which are characterized by positive feedback. Homeostasis may be thought of as all those relations which act to preserve a system by keeping it in a steady state during its existence. Homeorhesis may be thought of as all those relations which act to preserve not a steady state but a flow process which for an individual system follows a relatively fixed trajectory or, to use Waddington's (1957, 1968) term, a chreod; this is common in growth and development. With homeostasis, system pattern is preserved: homeostatic change is morphostatic; with homeorhesis, system pattern alters, usually becoming more differentiated and more complex: homeothetic change is morphodynamic. That a system may possess both these antithetical facets is something of a paradox; but they are complementary modes of regulation which enable a system to adapt to changes and challenges in its environment.

Many geographical systems are subject to homeostatic and homeorhetic change, though the latter is not controlled by a genetic template as it is in biological systems. In human geography, homeostatic and homeorhetic processes can be identified within the broader trend of societary and cultural evolution.[1] For example, the general pattern of change in communication network growth, as described by Taaffe *et al.* (1963), which recurs in many parts of the world, and some of the many diffusion processes studied by human geographers, for instance Hägerstrand (1967), seem homeorhetic in nature (Ray *et al.*, 1974). In biogeography, plant and animal succession is characterized by homeorhetic processes and leads to a climax vegetation endowed with an annually and diurnally

[1] Many systems, apart from being self-stabilizing, have a tendency to evolve. The terminology here needs some explanation: the word 'evolution' apart from having become somewhat emotive, has acquired too many nuances of meaning to be used unequivocally. Evolution can be understood in at least two ways (Mayr, 1970). Firstly, evolution can be taken in the literal sense of the unfolding or growth and development of an individual system. But, though during its development a system, owing to homeorhetic processes, may become more complex, the end product is not novel, merely a familiar, pre-existing form. For clarity, it might be better to refer to this process as 'homeorhesis' or simply 'development.' Secondly, evolution can be taken in the grander sense, as employed by Darwin for example for phylogenetic evolution—the derivation of all forms of life from a common simple ancestor—to describe the complexification (to use an ugly, rare, but precise word) of a system which produces a novel pattern at a so-called higher level. For clarity, it might be better to restrict the term evolution to this process of on-going complexification which would include the three main lines of evolution recognized by Huxley (1953), namely, inorganic evolution—the cosmic processes of matter (planetary evolution would be included with this); biological evolution—the evolution of plants and animals; and psychological evolution—the evolution of man and his cultures (following Popper (1974), this might be better thought of as the evolution of consciousness). In geography, Berry (1973) distinguished between evolutionary change and revolutionary change; but it is perhaps prudent to say that if evolution and revolution, as processes of complexification, differ at all, then they differ in tempo. Thus, the change from an agricultural society to an industrial society is an example of complexification which could be designated revolutionary, whereas the usually more gradual emergence of families, clans, villages, towns, cities, states, and nations could be deemed evolutionary.

[2] Stands are areas dominated by single-aged individuals of one species in which the individuals age roughly together and pass through four phases in the process: a regenerative phase, a building phase, a mature phase, and a degenerative phase.

[3] So-called stream-pattern evolution, as for instance observed by Morisawa (1964) who recorded the rapid integration and simultaneous expansion of gully networks, accompanied by tributary formation, extension, elimination, lengthening, and shortening, and the maintenance of an over-all equilibrium at all stages in the growth of small drainage basins, can be interpreted as the conjoint operation of homeorhetic and homeostatic processes. And so-called landscape evolution may be deemed a flow process which, for similar constraints in different areas, follows roughly the same course and is equifinal: the ultimate form of a landscape, if it be given time to develop, is a level plain. The intermediary states through which the landscape passes before reaching its final state are manifold. The path taken by a particular landscape is determined largely by the rate of land uplift, the rate of downcutting by streams, and the types of erosive process operative on the slopes, as well as an underlying geology. In the case of interfluvial slopes, if uplift should keep pace with downcutting then homeorhetic slope change will lead to a homeostatic (steady-state) slope form in which gain of potential energy by uplift will be balanced by erosion for as long as boundary conditions and the erosive processes remain fixed (Hirano, 1975). If uplift should cease then, under certain conditions, steady-state slopes may persist as the landscape gradually wears down; but as soon as downcutting stops the slopes will change irreversibly (Ahnert, 1971; Hirano, 1975). Used in this context, the terms 'homeostasis' and 'homeorhesis' are equivalent to, respectively, the terms 'timeless' and 'timebound' as used by Chorley (1966).

pulsing homeostasis. Sub-systems within a climax ecosystem, other than in individual organisms, may follow a homeorhetic path, the stand cycle recognized by Watt (1947) being an example of this.[2] In geomorphology, many 'evolutionary' processes are characteristically homeorhetic.[3]

5.2.4. *Deterministic process laws*

Many relations in systems can be expressed as laws. In physical systems these will be natural laws; in human systems these will be heuristic relations, usually physical laws applied to the human situation. Broadly speaking, four groups of deterministic equations are used to define system relations and states. These go by the mind-boggling names of balance equations, physical-chemical-state equations, phenomenological laws, and entropy-balance equations. For most applications in physical geography, these equations are firmly rooted in physical theory and mainly in that subject known as thermodynamics. For most applications in human geography, these equations are applied by analogy with physical systems. Thus, for instance, it has been found feasible to view the behaviour of a population as an aggregate of individual people in an analogous manner to viewing the behaviour of a gas as an aggregate of individual molecules; and to regard the attraction between two cities, as realized by the flow of information, people, or goods between them, analogously to the gravitational attractions of stars and planets, one to the other.

The façade of jargon which describes the four types of deterministic equations dissembles their rather straightforward meanings. Balance equations are a symbolic and precise way of showing that what goes into a system must be stored, come out, or be transformed into something else; matter, energy, and momentum cannot suddenly appear or disappear in the system in an unaccountable manner. In the case of mass transactions, this logical view is called the law of mass conservation; for energy transactions it is called the first law of thermodynamics which is a statement of energy conservation. The usual form of balance equations is

$$\frac{\text{change in state}}{\text{per unit time}} = \text{inputs} - \text{outputs}$$

where state could be a measure of mass, energy, or momentum. This balance rule was in fact used in drawing up the state transition function for the herbivore ecosystem (equation 5.14), but the equation was put in the form

$$\begin{matrix}\text{state at}\\ \text{time } t+1\end{matrix} = \begin{matrix}\text{state at}\\ \text{time } t\end{matrix} + (\text{inputs} - \text{outputs})\Delta t, \qquad (5.15)$$

Written in this style, general balance equations for mass, energy, and momentum look like this:

$$\begin{aligned}
&\text{mass at time } t+1 = \text{mass at time } t + \left(\text{mass inputs} - \text{mass outputs}\right)\Delta t \\
&\text{energy at time } t+1 = \text{energy at time } t + \left(\text{energy inputs} - \text{energy outputs}\right)\Delta t \\
&\text{momentum at time } t+1 = \text{momentum at time } t + \left(\text{incoming force} - \text{outgoing force}\right)\Delta t
\end{aligned} \tag{5.16}$$

This designation of balance equations has to be modified for spatial systems, such as a soil profile, a lake, a slope profile, and a river, in which system state depends upon location within the spatial domain of the system. The modification simply entails adding labels to the terms in the balance equations that, like map co-ordinates, indicate locations in space. A soil profile occupies a one-dimensional spatial domain, positions within which may be referred to depth increments along the axis running in the z direction (Figure 5.18). The heat balance for the portion of soil between points z and $z+1$ (Figure 5.18) could be given as

$$\text{heat stored in section } z \text{ at time } t+1 = \text{heat stored in section } z \text{ at time } t + \left(\text{heat inputs at point } z - \text{heat outputs at point } z+1\right) \tag{5.17}$$

Balance equations for different points in two-dimensional and three-dimensional spatial systems would include co-ordinates for the extra dimensions.

The inputs and outputs in the state transition function term of balance equations have to be defined by transport or process laws which, like the input and output expressions used in the state transition function for the herbivore system (equation 5.4), contain parameters. Transport laws in systems that are not studied in a spatial framework are said to have lumped parameters, whereas transport laws in spatial systems are said to

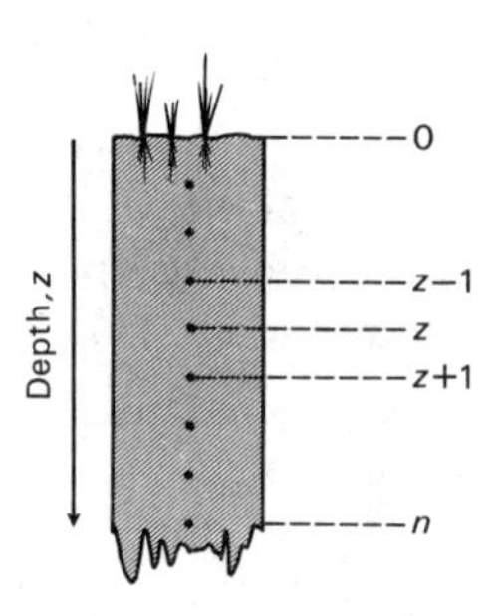

FIG. 5.18. Spatial reference system for a soil profile.

have distributed parameters (cf. Isermann, 1975). In the herbivore system, the loss of energy in respiration was set proportional to the standing crop (energy storage) of the herbivores (p. 82)

$$\begin{matrix}\text{output in} & & & \text{standing}\\ \text{herbivore} & = & \text{parameter} \times & \text{crop of}\\ \text{respiration} & & & \text{herbivores.}\end{matrix}$$

The 'herbivores' could be a mixed bag of animals—on the African savanna it would include elephants, zebras, gnus, and many others—each with its own respiratory characteristics. The parameter which determines the respiratory loss for the entire system has a value representative of all the herbivores in the area under study; it is an example of a lumped parameter. The equation itself is an example of a transport law. Indeed, the general relation

$$\text{output} = \text{parameter} \times \text{storage} \tag{5.18}$$

is applicable to many processes in non-living systems and processes in living systems where the output is passed to a non-living component of the system (as in respiration). Transport between living components of a system—herbivory and carnivory are cases in point—are thought to be better represented by a non-linear transport law which states that the flow is proportional to the product of the storages in the system components between which the flow is taking place.

In a spatial system, the transfer of mass, energy, and momentum from one place in the system to another depends upon the spatial pattern of both state variables and parameters, and the parameters are said to be distributed. Transport laws in spatial systems include a spatial term. The state transition function for the heat balance in a soil profile (equation 5.17) has heat-input and heat-output terms, both of which need to be defined by a heat transport law. A common assumption, supported by observational evidence is that heat flow through a point is proportional to the temperature gradient through the point. For heat flow in a soil profile we could write (see Figure 5.18)

$$\begin{matrix}\text{heat inputs} & = & \text{parameter} & \times & \text{temperature gradient}\\ \text{at point } z & & \text{at point } z & & \text{through point } z\end{matrix}$$

$$\begin{matrix}\text{heat outputs} & = & \text{parameter} & \times & \text{temperature gradient}\\ \text{at point } z+1 & & \text{at point } z+1 & & \text{through point } z+1.\end{matrix}$$

These equations conform to the heat-transport law known as Fourier's law of heat conduction. Fourier's law is an instance of a phenomenological equation, for, in common with all phenomenological laws, it describes a process which tends to equalize the value of a state variable, heat in this case, throughout the system—given no inputs nor outputs at the top and bottom of a soil profile, heat will move from hot areas to cool areas until the heat stored at all points is the same. The word 'phenomenological' is derived from a Greek word meaning 'to appear', and refers to the fact that changes in observable state variables, heat in Fourier's case, are described without trying to see what concomitant changes are occurring in atoms, molecules, or electrons. Other phenomenological laws include Fick's law of diffusion and Darcy's law of water flow through rock and soil.

Physical-chemical-state equations relate state variables one to another. The gas law of Boyle and Gay-Lussac is an example; it states

$$PV = RT$$

where P is pressure, V is volume, R is a gas constant, and T is temperature. Warntz (1973) gave an equation relating income density, D, area, A, population, P, and per capita income, T

$$DA = PT$$

and he noted the apparent similarity between his equation and the equation for a perfect gas: income density is like pressure; area is the counterpart of volume: total population is like the gas constant which is in fact proportional to the number of molecules in the gas; and per capita income is like temperature.

In physical geography, Leopold and Langbein (1962) noticed that a heuristic analogy could be set up between landscape variables and thermodynamic variables; they showed that as a thermodynamic field is described by temperature and heat, so a landscape is analogously described by the height of the land above a base line and mass. Scheidegger (1970) developed these ideas and, for example, gave an analogue for landscape pressure derived from the ideal gas law:

$$p_{\text{landscape}} = \text{constant}\,\frac{h}{A}$$

where A is the area under consideration. If a slope profile is considered this becomes

$$p_{\text{landscape}} = \text{constant}\,\frac{h}{L}$$

where L is the length of the profile.

5.3. Stochastic relations

Many relations in systems, especially those in human geography, cannot as yet be written as exact functional relations because of a lack of theory.

In the absence of laws to define system state, the assumption can be made that system state at a given time is influenced by, though not rigidly controlled by, previous system states. Thus the present economic and demographic landscape of England may be assumed, owing to processes such as inertia, momentum, and continuity, to be influenced by the economic demographic landscape of say a decade ago. The relation between one state and a previous one may be expressed as a Markov chain. More sophisticated models have been devised which take into account the relations between states in geographical space as well as in time and assume that the value of a state variable in one region at one time is dependent upon of the value of the state variable in the same region and other regions at previous times; these follow two main approaches. The first one is that taken by Hägerstrand (1967) who, using what are known as Monte Carlo techniques, modelled the spread of innovations through a region by making a number of behavioural assumptions about contacts between people. Relations in Monte Carlo models are in essence seen as random movements of a system component, say population or ideas, between different regions during discrete time steps. The second of the two approaches is to regard the real world as a black box and set up a statistical equation which relates the distribution of, say, population at one time to the distribution of population at another time. Two main kinds of equation are used for this purpose: one kind are state transition functions and the other kind are auto-regressive, moving-average, or space-time regression equations (see Bennett, 1975; Martin and Oeppen, 1975). This statistical approach, which is no more than a sophisticated form of extrapolation, has been found valuable in making regional forecasts; the details of it are too advanced for this book and the interested and determined reader should consult Haggett *et al.* (1977) for a general account, and Hepple (1974) for a technical account. We shall here consider relations as Markov chains and in Monte Carlo models.

5.3.1. *Relations as Markov chains*

In Markov chains, the change from one system state to another is assumed to be effected by random processes. This idea was first posited by the Russian mathematician Markov and is now found in geographical literature. A Markov chain can be thought of as a series of transitions between different system states, the transitions, which are expressed as probabilities, depending only on preceding states. If the present state of a system depends on the state at the immediately preceding time only, the Markov chain extends back over one time step and is a first-order Markov chain. If the dependence of state extends back to even earlier states a higher-order Markov chain is formed.

The dependence of one state on another is expressed in transition probabilities. We may put the notions more formally for a hypothetical example.

Take the case of daily weather change. Dividing weather type into three groups or states—fine, fair, and foul, we may keep a record of the number of days in, say, one year the weather passes from one state to another. The record for one year is shown in Table 5.3 which may be converted into a table of transition probabilities between weather types. For instance, a fine day of weather follows a day of fair weather 66 times out of the 165 times that fair weather was recorded during the year: the transition probability from fair to fine is $66/165 = 0.4$. With all transition probabilities calculated in like manner we arrive at Table 5.4.

Table 5.3
Daily weather changes over one year

		to			
		Fine	*Fair*	*Foul*	*Row totals*
	Fine	48	32	80	160
from	Fair	66	66	33	165
	Foul	12	24	4	40
					$\Sigma = 365$

The transition probabilities in Table 5.4 can be represented as a square array of numbers or matrix in which rows and columns correspond to system states. Calling this matrix **P**, we have

$$\mathbf{P} = \begin{array}{c} \\ \text{Fine} \\ \text{Fair} \\ \text{Foul} \end{array} \begin{array}{c} \begin{array}{ccc} \textit{Fine} & \textit{Fair} & \textit{Foul} \end{array} \\ \begin{pmatrix} 0.3 & 0.2 & 0.5 \\ 0.4 & 0.4 & 0.2 \\ 0.3 & 0.6 & 0.1 \end{pmatrix} \end{array}.$$

Or, in the general case for a three-state system

$$\mathbf{P} = \begin{array}{c} \\ x_1 \\ x_2 \\ x_3 \end{array} \begin{array}{c} \begin{array}{ccc} x_1 & x_2 & x_3 \end{array} \\ \begin{pmatrix} p_{11} & p_{12} & p_{13} \\ p_{21} & p_{22} & p_{23} \\ p_{31} & p_{32} & p_{33} \end{pmatrix} \end{array}$$

where the ps are transition probabilities between states x_1, x_2, and x_3. Note that a transition probability of 1.0 means that once entered, that state will not change and is known as an absorbing state. The information

Table 5.4
Transition probabilities between weather types

		to			
		Fine	*Fair*	*Foul*	*Row totals*
	Fine	0.3	0.2	0.5	1.0
from	Fair	0.4	0.4	0.2	1.0
	Foul	0.3	0.6	0.1	1.0

in a transition probability matrix, such as the daily weather change one, can be depicted in diagrammatic form (Figure 5.19).

5.3.2. *Relations in Monte Carlo models*

Problems in which movement of system components—people, money information, ideas, plant seeds, viruses, atomic particles, or whatever—cannot be represented by a tractable deterministic relation, may usually be formulated in stochastic terms as a Monte Carlo model. In essence, stochastic relations in a Monte Carlo model are defined as a random variable field, sampling from which enables the movement of system components to be traced and the corresponding change of system state to be recorded. For instance, in Hägerstrand's (1967) classic application of the Monte Carlo technique to the diffusion of new agriculture practices, such as tuberculosis control in cattle, and pasture-improvement subsidies, a mean information field is set up which defines the likelihood of a farmer being told of an agriculture innovation by a farmer who already knows of it (the teller) and whose farm is at centre of the mean information field (Figure 7.15). As can be seen, the mean information field is a square probability distribution which is constructed in three stages, each stage incorporating certain assumptions. In the first stage, an unweighted mean

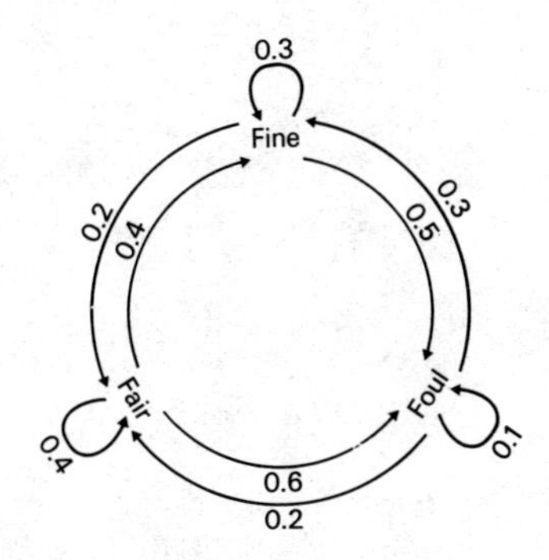

FIG. 5.19. State changes as transition probabilities.

information field is set up; the set of values in this field are the probabilities that a farmer located in the central cell will tell a farmer located in any other cell about the innovation. The neighbourhood effect—the greater likelihood of a farmer telling other farmers in the nearer cells—is incorporated in this field. In the second stage, the unweighted information field is modified by allowing for the actual distribution of farmers in the area over which the field is laid. And in the third stage, modifications are made in accordance with the assumed effect of environmental barriers—lakes, rivers, and so forth—on the contact probabilities. The result is the mean information field. It is very important in this and other examples of stochastic relations in Monte Carlo models that the form of probability field used should be logical and based on a theory of some kind. For instance, in Hägerstrand study of the spread of pasture improvement subsidies, six justifiable assumptions concerning relations in the system were made: a limited number of farmers have the information to start with; the innovation—pasture improvement—is accepted at once when heard of; the information spreads only through pairwise contacts between farmers; contact between farmers is made at discrete time intervals; at each step, every farmer who knows about the innovation tells just one other farmer who may or may not already know about it; the probability of a farmer who knows about the innovation contacting a farmer who does not or a farmer who does is some function of distance away from the teller.

6 Flow Models

True wit is nature to advantage dressed,
What oft was thought but ne'er so well expressed.
POPE, *An Essay on Criticism*

6.1. Constructing compartment models

EARLIER in the book it was seen that a system could be regarded as a set of linked components. One method of viewing such a system is to regard the components as stores of matter or energy. The amount of matter or energy in the store may be measured and is a state variable. Take as an example the carbon cycle in the Aleutian ecosystem which we looked at in chapter 2; this system can be represented by nine state variables—the atmosphere, land plants, terrestrial dead organic matter, man, phytoplankton, zooplankton and marine animals, marine dead organic matter, surface water, and deep sea—each of which can be thought of as a compartment which stores carbon (Figure 6.1). The relations between the system components take the form of inputs and outputs of carbon as it flows through the Aleutian ecosystem; the direction and magnitude of these flows is shown in Figure 6.1, which is a compartment model in diagrammatic form. Notice that the model of the Aleutian ecosystem, like the other models that will be discussed in this chapter, is aspatial in as much as the locational and spatial facets of compartments are not explicitly considered.

6.1.1. *Mathematical formulation: from flow diagrams to equations*

The information in Figure 6.1 can be converted into a potentially more powerful mathematical format. For any one flow of carbon, a source or donor compartment from which the flow emanates, and a terminal or receptor compartment, by which the flow is received, may be defined. The flow between the atmosphere and land plants, because it is from state variable x_1 to state variable x_2, may be labelled F_{12}. All other flows can be labelled using the same logic (Figure 6.1). The value of a state variable, say man, x_4, at time $t+1$, $x_{4,t+1}$, depends on the value of the state variable at time t, $x_{4,t}$, the time interval between t and $t+1$, Δt, and a state transition function which in this example will be the difference between incoming and outgoing flows of carbon. The incoming flows of carbon are F_{24} and F_{64}, and the outgoing flows are F_{41} and F_{43}, so we may write a balance equation for carbon storage in man as

$$x_{4,t+1} = x_{4,t} + (F_{24} + F_{64} - F_{41} - F_{43})\Delta t.$$

Bear in mind that this equation can also be expressed as

$$\frac{\Delta x_4}{\Delta t} = F_{24} + F_{64} - F_{41} - F_{43}$$

time rate of carbon storage in man	incoming flows of carbon	outgoing flows of carbon
	state transition function	

and commonly is in books and papers dealing with systems analysis of ecosystems. An equation can be drawn up to describe the change in each state variable; the result is a set of nine equations which look like this:

$$x_{1,t+1} x_{1,t} + (F_{21} + F_{31} + F_{41} + F_{81} - F_{12} - F_{18}) \Delta t$$

$$x_{2,t+1} = x_{2,t} + (F_{12} - F_{21} - F_{23} - F_{24}) \Delta t$$

$$x_{3,t+1} = x_{3,t} + (F_{23} + F_{43} - F_{31}) \Delta t$$

$$x_{4,t+1} = x_{4,t} + (F_{24} + F_{64} - F_{41} - F_{43}) \Delta t$$

$$x_{5,t+1} = x_{5,t} + (F_{85} - F_{56} - F_{57}) \Delta t \tag{6.1}$$

$$x_{6,t+1} = x_{6,t} + (F_{56} - F_{64} - F_{67}) \Delta t$$

$$x_{7,t+1} = x_{7,t} + (F_{57} + F_{67} - F_{78} - F_{79}) \Delta t$$

$$x_{8,t+1} = x_{8,t} + (F_{18} + F_{78} + F_{98} - F_{81} - F_{85} - F_{89}) \Delta t$$

$$x_{9,t+1} = x_{9,t} + (F_{79} + F_{89} - F_{98}) \Delta t.$$

The flows, or fluxes as they are usually termed, can be defined in terms of the state variables. For many transfer processes, the flux between two compartments is proportional to the amount stored in the donor compartment

$$F_{ij} = \lambda_{ij} x_i \tag{6.2}$$

where λ_{ij} is a constant called a transfer coefficient. In the Aleutian ecosystem, the flux between the atmosphere and surface water would be defined as

$$F_{18} = \lambda_{18} x_1.$$

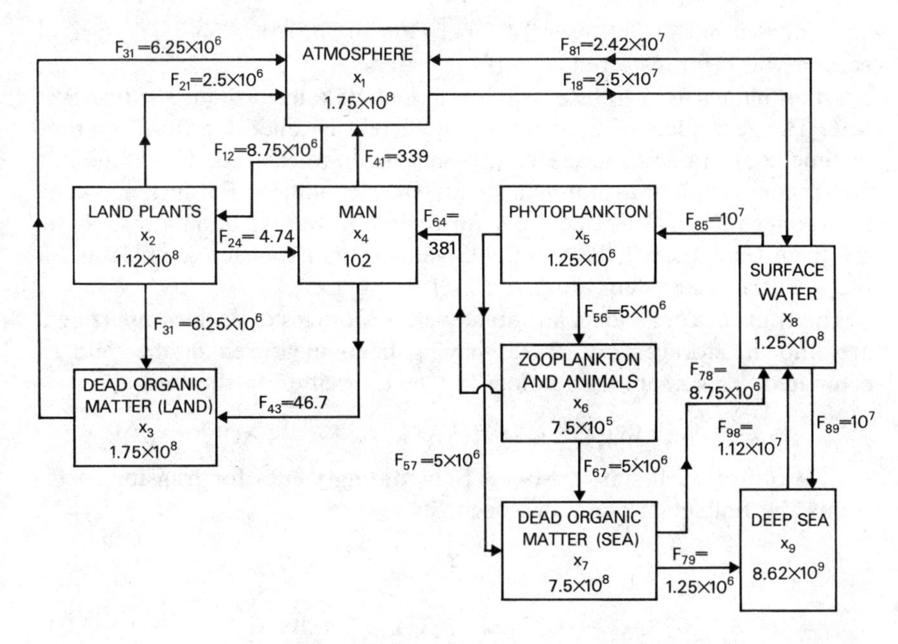

FIG. 6.1. Carbon model of the Aleutian ecosystem. The values in each compartment represent metric tons (10^3) of carbon at steady state. Numbers on the arrows represent the flux of carbon (metric tons per year) between compartments. After J. M. Hett and R. V. O'Neill (1974). See Figure 2.1. (p. 22).

In practice, the flux and storage in equation 6.2 are usually known, having been measured in the field, as they have for the Aleutian ecosystem (Figure 6.1), but the transfer coefficients need calculating. Rearranging equation 6.2 gives

$$\lambda_{ij} = \frac{F_{ij}}{x_i} \tag{6.3}$$

which thus defines the value of the transfer coefficient. In the case of the carbon flux between atmosphere and surface waters

$$\lambda_{18} = \frac{F_{18}}{x_1}$$

and substituting values from Figure 6.1 this gives

$$\lambda_{18} = \frac{2.5\times 10^7}{1.75\times 10^8}$$

$$= 0.143 \text{ per year.}$$

This means that in one year 14.3 per cent of the atmospherical store of carbon will be transferred to surface waters.

Other names for a transfer coefficient include rate constant and turnover rate. The reciprocal of the transfer coefficient is called the turnover time or time constant and represents the time taken to replace the storage in a donor compartment assuming a steady-state condition. For instance, with a turnover rate of 0.143 per year for carbon in the Aleutian atmosphere, the turnover time is $1/0.143 = 6.993$ years. Other transfer coefficients for the Aleutian ecosystem are given in Table 6.1.

The transfer coefficients and steady-state storages of each compartment are known, storages and fluxes having been measured in the field or estimated. This enables equations 6.1 to be written in the form

$$x_{1,t+1} = x_{1,t} + (\lambda_{21}x_2 + \lambda_{31}x_3 + \lambda_{41}x_4 + \lambda_{81}x_8 - \lambda_{12}x_1 - \lambda_{18}x_1)\Delta t$$

and so on for each state variable. Substituting values for transfer coefficients the full set of equations becomes

$$\begin{aligned}
x_{1,t+1} &= x_{1,t} + (0.0223x_2 + 0.0357x_3 + 3.32x_4 + 0.194x_8 - 0.05x_1 - 0.143x_1) \\
x_{2,t+1} &= x_{2,t} + (0.05x_1 - 0.0223x_2 - 0.0558x_2 - 4.21 \times 10^{-8}x_2)\Delta t \\
x_{3,t+1} &= x_{3,t} + (0.0558x_2 + 0.458x_4 - 0.0357x_3)\Delta t \\
x_{4,t+1} &= x_{4,t} + (4.21 - 10^{-8}x_2 + 5.08 \times 10^{-4}x_6 - 3.32x_4 - 0.458x_4)\Delta t \\
x_{5,t+1} &= x_{5,t} + (0.08x_8 - 4.00x_5 - 4.00x_5)\Delta t \qquad (6.4) \\
x_{6,t+1} &= x_{6,t} + (4.00x_5 - 5.08 \times 10^{-4}x_6 - 6.67x_6)\Delta t \\
x_{7,t+1} &= x_{7,t} + (4.00x_5 + 6.67x_6 - 0.0113x_7 - 1.67 \times 10^{-3}x_7)\Delta t \\
x_{8,t+1} &= x_{8,t} + (0.143x_1 + 0.0113x_7 + 1.3 \times 10^{-3}x_9 - 0.194x_8 \\
&\qquad - 0.08x_5 - 0.08x_8 \\
x_{9,t+1} &= x_{9,t} + (1.67 \times 10^{-3}x_7 + 0.08x_8 - 1.3 \times 10^{-3}x_9)\Delta t
\end{aligned}$$

and this is a fully calibrated, mathematical model of the Aleutian ecosystem.

6.1.2. *Making the model work*

Equations 6.4 are the mathematical formulation of the model; they contain all the information regarding system behaviour. Given initial values for the state variables, the equations can be solved to show how the state variables change with time. Several methods are available for doing this and we shall consider the most basic one, known as the Euler or Rectangular method, applying it, not to the relatively complex set of

Table 6.1
Transfer coefficients, λ_{ij}, for carbon model in the Aleutian ecosystem

		from state variables								
		x_1	x_2	x_3	x_4	x_5	x_6	x_7	x_8	x_9
	x_1	—	0.0223	0.0357	3.32				0.194	
	x_2	0.05	—							
	x_3		0.0558	—	0.458					
	x_4		4.21×10^{-8}		—		5.08×10^{-4}			
to	x_5					—			0.08	
	x_6					4.00	—			
	x_7					4.00	6.67	—		
	x_8	0.143						0.0113	—	1.3×10^{-3}
	x_9							1.67×10^{-3}	0.08	

From Hett, J. M. and O'Neill, R. V. (1974). Systems analysis of the Aleut ecosystem. *Arctic Anthropology*, **XI. 1.** 31–40. Reprinted by permission of, and copyright © 1974, The Board of Regents of the University of Wisconsin System.

equations describing the dynamics of the Aleutian ecosystem, but to the simple equation

$$x_{t+1} = x_t - (\lambda x_t)\Delta t \tag{6.5}$$

which could be used to describe, among other things, the decay of a tree which has fallen on a forest floor. In the example, x_{t+1} is the mass of the tree at time $t+1$; x_t is the mass of the tree at time t; λ is a transfer coefficient (in this case the fractional decay rate); and Δt is a time interval. To obtain a solution to equation 6.5, three things must be specified: the mass of the tree when first it fell—this is the initial state of the system, the value of x at time $t=0$, denoted by x_0, which we shall set at 100 tonnes; the transfer coefficient—we shall assume this has been established in the field as 0.5 per year (that is, 50 per cent per year); and the length of the time interval, Δt—this we shall set at one year, though we shall see later the effect the length of the interval has on the results. When the tree has just fallen, that is when $t=0$, equation 6.5 may be written

$$x_1 = x_0 - (\lambda x_0)\Delta t.$$

Substituting the known values into the right-hand side of this, we get

$$\begin{aligned} x_1 &= 100 - (0.5 \times 100)1.0 \\ &= 50 \text{ tonnes.} \end{aligned}$$

So a year after it first fell, the tree has a mass of 50 tonnes. The mass of the tree after a second year has passed may now be computed by putting

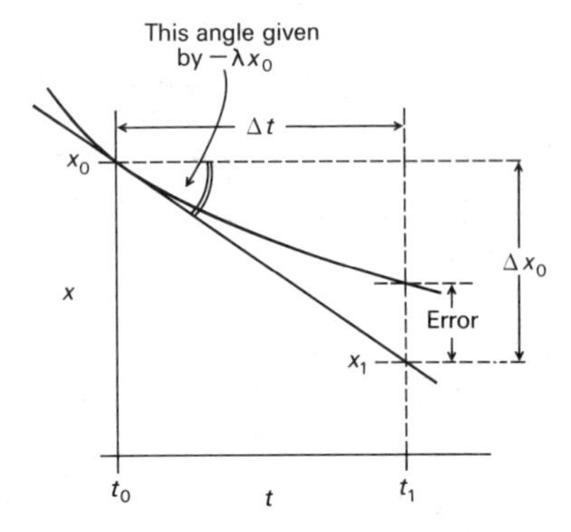

FIG. 6.2. The decay of a fallen tree in a tropical rain forest. Notice that the error in the numerical solution increases with time and with increasing size of the time interval Δt.

the value of x_1 into the equation

$$x_2 = x_1 - (\lambda x_1)\Delta t$$

which produces

$$x_2 = 50 - (0.5 \times 50)1.0$$
$$= 25 \text{ tonnes.}$$

This procedure can be repeated for any desired number of time steps. An error is involved in this method (Figure 6.2), and the greater the number of time steps used, the greater the error (Figure 6.3). The error can be controlled by making the time interval Δt as short as possible; but even with this safeguard, the solution at times far removed from $t = 0$ will be appreciably in error and the solution may become unstable giving nonsensical results. Far more accurate alternatives to the Euler method are described by Patten (1971) and Gunn (1971). Several computer

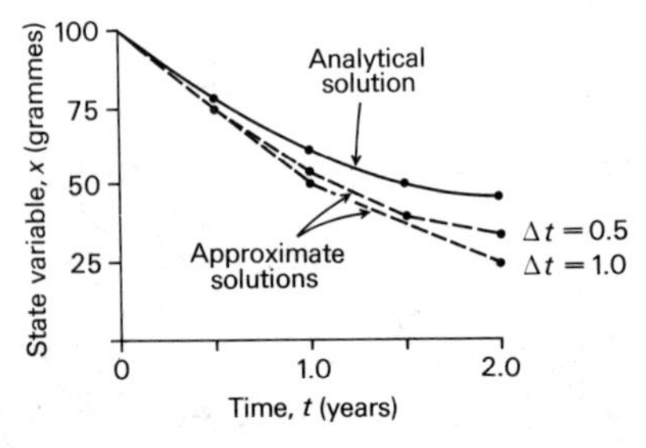

FIG. 6.3. Analytical and numerical solutions to equation $\Delta x/\Delta t = -\lambda x$.

languages, such as DYNAMO, CSMP, and DAREP have been developed specifically for solving sets of system equations, such as equations 6.4, and thus simulating the behaviour of systems.

6.1.3. *Using the model*

Compartment models may be used in two main ways: in the solution of difference equations describing the behaviour of the system; and in the identification of soft-spots in the system by studying the stability of the solutions (Tomović, 1963). The solution of the equations is sometimes valuable in predicting future states of a system. Large computers can crunch through ghastly looking equations with relative ease and churn out sheets of beautiful output showing changes in state variables over time. Unfortunately, these results are deceptively precise and often difficult to test against the real-world situation. As was noted earlier (p. 26) two methods of testing the results are available. A particular response can be simulated on a computer for which actual field data are available. If agreement between the model results and reality is poor, the model can be adjusted, by changing a few flow rates for instance, until computer behaviour and actual behaviour are in accord. Alternatively, postdictions can be made, that is, the system behaviour is computed over a time period for which data on actual behaviour are available. The behaviour of the British economy, for instance, between 1960 and 1970 could be simulated and the results matched against actual economic changes in that decade. Should the model work well, predictions beyond 1970 would seem in order. In practice, long-term simulations into the future, especially in the case of economic and social systems, are noted for their inaccuracy of detail; this is mainly caused by unforeseen changes in external variables—the rise in oil prices for example. Predictions of change in ecological systems have fared much better as we shall see later in this chapter. Compartment-type models have been responsible for bringing to light the complex adjustments systems make in response to changes in input or changes in internal structure, so dispelling the notion that system components are linked by simple cause and effect sequences—*a* causes *b*, *b* then causes *c*, and so forth—and replacing it by the idea of counter-intuitive behaviour, so called because a simple change in a system which is thought will produce a single effect often leads to unexpected effects in other parts of the system. Thus fertilizing field crops accidentally boosts algal growth in streams and reduces fish populations; pesticides cause an increase in pest populations by knocking out natural predators; and so on. In short, the results of a computer simulation showing future trends, notably in the case of systems of interest to human geographers, should not be taken as trustworthy predictions of what the system will in fact do; instead they indicate the types of behaviour which, given the sheer complexity of interaction within the system would be impossible to establish intuitively (Waddington, 1977, p. 104).

Studying the stability of solutions to systems models enables soft-spots in the system to be located. The process by which this is done is called sensitivity analysis. The technicalities of sensitivity analysis are a bit irksome but in essence the procedure is to run a set of experiments on a computer, the value of a system parameter or state variable being changed by a fixed amount in each experiment. The parameters which have greatest effect in system behaviour can thus be identified as well as the state variables which are most sensitive to changes in parameters or flows. Work of this kind on relatively simple ecosystems suggests that the most significant changes in behaviour are brought about by changes in the organizational structure of the relations between system components, the values of the state variables themselves being of lesser importance (Cooper, 1969; Jordan and Kline, 1972).

6.1.4. *Input–output flow analysis*

Much can be learnt about a system by studying the nature of material or energy flow through it. In particular, it is possible to derive flow measures, notably useful in comparing cycles in two or more systems, which represent the efficiency of system components and an entire system in directing flows so that as much matter or energy as possible is recirculated within the system rather than passing straight through. Flow analysis, or input-output flow analysis to give it its full title, first used by Leontief (1966) in studying economic systems, has recently been adapted to analysing flows in ecosystems (see Patten *et al.*, 1976). Input–output flow analysis requires a fair bit of mathematical expertise and in this book we shall illustrate its use by means of the simple example of an ecosystem studies by Rykiel and Kuenzel (1971)—the wolves of Isle Royale ecosystem.

The Isle Royale ecosystem, for our purposes, consists of three system components (state variables), viz. plants, moose, and wolves. The steady-state storages and flows of energy in the system are shown in Figure 6.4. The first step in the input-output flow analysis of the system is the construction of a production matrix, similar to the transaction matrix in economic systems (see Chenery and Clark, 1959; Wilson, 1974), which we shall denote by the letter **P**. This matrix tabulates all the flows to, within, and from the system. As can be seen in Table 6.2, which is the production matrix for the Isle Royale ecosystem, 19.4 kcal/m^2/yr pass from plants, x_1, to moose, x_2; and 0.085 kcal/m^2/yr pass from moose, x_2, to wolves, x_3; plants receive 7500 kcal/m^2/yr as solar energy and this value is entered in the first row under the column headed inflow; plants lose a total of 7480.6 kcal/m^2/yr as respiration, death, and to other

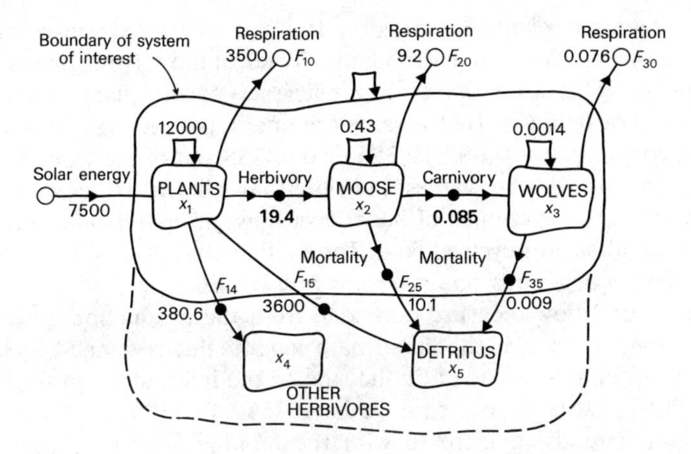

FIG. 6.4. The Isle Royale ecosystem. Storage is measured as kcal/m^2; flow as kcal/m^2/yr.

herbivores, and this total is entered in the first column of the outflow row; similarly, moose lose a total of 19.3 kcal/m^2/yr and wolves 0.085 kcal/m^2/yr, which values are placed in columns 2 and 3 respectively of the outflow row. Individual rows and columns are then summed. For a system in a steady state, such as the Isle Royale one, the sum of the column totals will equal the sum of the row totals and is termed the total system throughflow; in our example total system throughflow is 7519.49 kcal/m^2/yr.

A couple of flow measures may be derived from the production matrix. The first, the component cycling efficiency, may be calculated for each state variable by expressing the endogenous outflow—the outflow from a component which is passed on to other components within the system—as a ratio of the total outflow—which includes flows which leave the system.

Table 6.2
*Production matrix, **P**, for Isle Royale ecosystem*

F_i \ F_j	from x_1	x_2	x_3	*inflow*	*outflow*	*row total*
to x_1	0	0	0	7500	0	7500.0
x_2	19.4	0	0	0	0	19.4
x_3	0	0.085	0	0	0	0.085
inflow	0	0	0	0	0	0
outflow	7480.6	19.3	0.085	0	0	7499.985
column total	7500.0	19.385	0.085	7500	0	7519.49
cycling efficiency	0.00258	0.00438	0.0			

In our example, plants, x_1, pass on 19.4 kcal/m²/yr to moose and this is endogenous outflow from the plants. Total outflow from plants is 7500 kcal/m²/yr. Therefore the cycling efficiency of the plant component is 19.4/7500 or 0.00258. By the same argument, the cycling efficiency of the moose component is 0.085/19.385 or 0.00438 and of the wolf component is 0/0.085 or 0. The values indicate that all the state variables are inefficient in so far as most of the energy flow through them is fed into the environment as non-cyclical flow. The cycling efficiency of a state variable which has endogenous outflows only will be 1.0.

The second flow measure derivable from the production matrix, for a steady-state system, is the mean path length; this is defined as the total system throughflow divided by the sum of the inflows (or outflows) in the production matrix. In our case we have 7519.49/7500 or 1.00259. Mean path length tends to increase with the number of system components, greater throughflow, and more feedback (endogenous flow).

Further input–output flow analysis requires some numerical juggling with the production matrix, and involves both inflow analysis and outflow analysis. We shall consider inflow analysis; the procedures for outflow analysis are the same. The first stage in inflow analysis is the construction

Table 6.3
Matrices used in inflow analysis

(a) Fractional inflow matrix, $\mathbf{Q}'$

F_i \ F_j	from x_1	from x_2	from x_3	F_{10}
to x_1	0	0	0	1
to x_2	1	0	0	0
to x_3	0	1	0	0
F_{10}	0	0	0	0

(b) $(\mathbf{I}-\mathbf{Q}')$ matrix

F_i \ F_j	from x_1	from x_2	from x_3	F_{10}
to x_1	1	0	0	−1
to x_2	−1	1	0	0
to x_3	0	−1	1	0
F_{10}	0	0	0	1

(c) *Transitive closure inflow matrix,* $(\mathbf{I}-\mathbf{Q}')^{-1}$

F_i \ F_j	from x_1	from x_2	from x_3	F_{10}	outflow path length
to x_1	1	0	0	1	2
to x_2	1	1	0	1	3
to x_3	1	1	1	1	4
F_{10}	0	0	0	1	1

(d) *Throughflow matrix,* $\mathbf{T}$

F_i \ F_j	from x_1	from x_2	from x_3	F_{10}
to x_1	7480.6	0	0	7480.6
to x_2	19.3	19.3	0	19.3
to x_3	0.085	0.085	0.085	0.085
F_{10}	0	0	0	0
T_j	7499.985	19.385	0.085	
inflow Σ				7499.985

of a fractional inflow matrix, designated **Q′** (Table 6.3a). Elements in this matrix express the inflow from one state variable to another as a fraction of the throughflow in a particular state variable. Thus, in Table 6.3a the value 1 in position $\mathbf{Q}'_{14}$ is the inputs to the plants, 7500 kcal/m^2/yr, divided by the total throughflow in the plants which is also 7500 kcal/m^2/yr (shown as the row total). Notice in the Isle Royale example the elements of the fractional inflow matrix are either 1s or 0s: in more complex cases they would be fractions. The second stage of inflow analysis involves tricky matrix manipulations. The fractional inflow matrix is subtracted from an identity matrix, **I**, of the same size as **Q′** (in our case 4×4) and which by definition has 1s as the elements in the main diagonal and 0s elsewhere; this gives an (**I–Q′**) matrix (Table 6.3b). The (**I–Q′**) matrix is then inverted (see Sumner, 1978, chapter 4) to give what is termed a transitive closure inflow matrix, $(\mathbf{I}-\mathbf{Q}')^{-1}$ (Table 6.3c). The row totals of the transitive closure inflow matrix are termed the outflow path lengths; the elements of the matrix represent all direct and indirect flows in the systems. Premultiplying the transitive closurel inflow matrix by the outflow row from the production matrix (Table 6.3d) (put into diagonal form) produces a matrix of throughflows, **T**. The throughflow matrix can be interpreted in conjunction with the production matrix as follows. Outflow from plants is 7480.6 kcal/m^2/yr; this requires a contribution of 7480.6 kcal/m^2/yr to plant inflow, the total value of which from the production matrix is 7500 kcal/m^2/yr. Outflow of 19.3 kcal/m^2/yr from the moose requires 19.3 units from F_{10}, seen as a contribution of 19.3 units to the throughflows of plants and moose. Outflow of 0.085 kcal/m^2/yr from the wolves requires 0.085 kcal/m^2/yr from F_{10}, seen as a contribution of 0.085 kcal/m^2/yr to each of the throughflows of plants, moose, and wolves.

The transitive closure inflow matrix can be used to measure the efficiency of the system in cycling flows. For instance, applying a somewhat involved algorithm, the efficiency of a system component in feeding throughflow into feedback loops can be worked out; this measure is called loop-cycling efficiency. The weighted sum of loop-cycling efficiencies of all system components measures the system-cycling efficiency, the value of which will approach 0 for a system in which little material is cycled within the system and will approach 1 for a system in which most of the flow is cycled within the system. Also from the transitive inflow closure matrix may be derived a cycling index which measures, in effect, the ratio between feedback and non-feedback cycling to the system. So a cycling index of 0.25 would mean that, of the energy or material flowing through the system, a quarter as much is cycled as passes straight through. Some flow measures for three different ecosystems are listed in Table 6.4.

Table 6.4
Some flow measures in contrasting ecosystems

Ecosystem	*Total system throughflow*	*Cycling efficiency*	*Cycling index*
Marine*	549.3 $gC/m^2/yr$	0.116	0.24
Cold water spring	30626.5 $kcal/m^2/yr$	0.138	0.16
Tropical rain forest	71.8 $gN/m^2/yr$	0.492	1.78

* Detritus feeders in a benthic ecosystem.
Adapted from Patten *et al.* (1976).

6.2. Ecological models

6.2.1. *Mineral cycles: radioactive materials in ecosystems*

In order to investigate the long-term dynamics of radioactive isotopes in ecosystems, which resulted from the atmospherical testing of nuclear devices in the late 1950s and early 1960s, Jordan *et al.* (1973) carried out a long-term investigation of the cycling of radioactive and stable isotopes in a Puerto Rican tropical rain forest. They carried out a systems analysis of the dynamics of the stable isotopes of strontium and manganese and then used a computer model based on this analysis to predict the dynamics of radioactive isotopes of strontium and manganese in the ecosystem. The tropical rain forest was modelled as four compartments: canopy, litter, wood, and soil (Figure 6.5). Steady-state fluxes and storages of the stable manganese and strontium were measured in the field (see Jordan *et al.* 1973 for methods) and are shown in Figure 6.5.

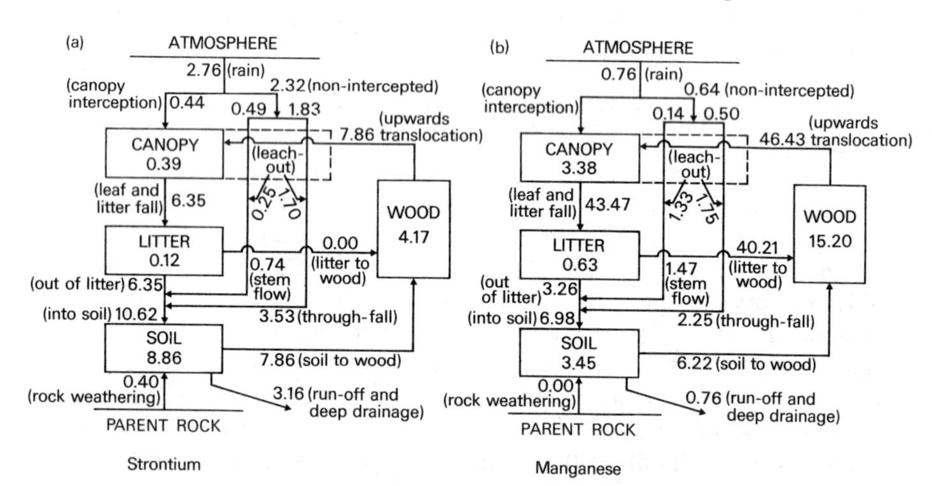

FIG. 6.5. Observed flows and storage of manganese and strontium in a Puerto Rican tropical rain forest. From C. F. Jordan *et al.* (1973). Reproduced from *Health Physics*, **24** (1973), 477–89, by permission of the Health Physics Society.

The model is summarized in the following balance equation in which the term in brackets is the state transition function:

$$x_{j,t+1} = x_{j,t} + \left(\sum_{\substack{i=0 \\ i \neq j}}^{4} \lambda_{ij} x_{i,t} - \sum_{\substack{i=0 \\ i \neq j}}^{4} \lambda_{ji} x_{j,t} - \lambda_r x_{j,t} \right) \Delta t, \qquad j = 1,2,3,4$$

$x_{j,t+1}$	$x_{j,t}$	$\sum \lambda_{ij} x_{i,t}$	$\sum \lambda_{ji} x_{j,t}$	$\lambda_r x_{j,t}$	Δt
material stored in compartment j at time $t+1$	material stored in compartment j at time t	inputs to j from all other compartments plus the environment	outputs from j to all other compartments and the environment	radio-active decay in compartment j	time

where $j = 1, 2, 3, 4$ are the compartments of the ecosystem (canopy, litter, wood, soil), $i = 0$ is the environment; x_j is the amount of radioactive material per unit area in compartment j; λ_{ij} is a transfer coefficient for the radioactive material from compartment i to compartment j; and λ_r is a suitable radio-active decay constant. The transfer coefficients were calculated from the equation

$$\lambda_{ij} = \frac{F_{ij}}{x_i}$$

where F is the steady-state flux of material out of a compartment and x_i is the steady-state storage of the donor compartment. Thus, in the manganese cycle, the balance equation for manganese in the wood compartment, ignoring the radioactive decay term, is written

$$x_{\text{wood},t+1} = x_{\text{wood},t} + \left(\underbrace{\frac{6.22 \times 10^{-3}}{3.45} x_{\text{soil},t}}_{\text{input from soil}} + \underbrace{\frac{40.20 \times 10^{-3}}{0.63} x_{\text{litter},t}}_{\text{input from litter}} - \underbrace{\frac{46.43 \times 10^{-3}}{15.20} x_{\text{wood},t}}_{\text{output}} \right) \Delta t.$$

Similar equations were drawn up for the other three compartments. Using a digital computer to solve the set of four simultaneous difference equations, two types of analysis were carried out. In the first analysis the input of radioactive isotopes was adjusted to resemble the actual radioactive fall-out conditions at the Puerto Rican site which resulted from the atmospherical testing of nuclear weapons. In the second analysis, an instantaneous injection of radioactive strontium, ^{90}Sr, and manganese, ^{54}Mn, was used because this is of practical interest if thermonuclear devices were to be used for the excavation of harbours or canals in the tropics. The predictions using observed fall-out patterns as input are

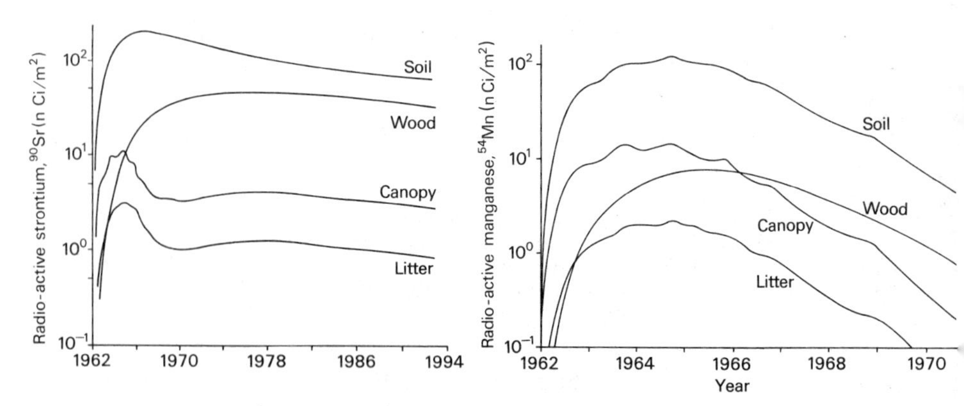

FIG. 6.6. Predicted storages of manganese and strontium in a Puerto Rican tropical rain forest using observed radioactive fall-out patterns as input. From C. F. Jordan *et al.* (1973). Reproduced from *Health Physics*, **24** (1973), 477–89, by permission of the Health Physics Society.

shown in Figure 6.6. In the case of the strontium-90 model, the compartments have the following maxima of ^{90}Sr activity: canopy and litter, 1963; soil 1967; wood, 1976. A secondary maximum occurs in the canopy and litter, owing to recycling from the soil, in 1977. Although after the initial input the soil shows the greatest amount of ^{90}Sr, followed by the wood, canopy, and litter, the concentration of ^{90}Sr (that is, the mass of ^{90}Sr divided by the mass of the compartment), is greatest in the canopy, followed by the litter, wood, and soil. In the case of ^{54}Mn, the peaks for the soil, canopy, and litter are roughly the same as for ^{90}Sr; but the wood

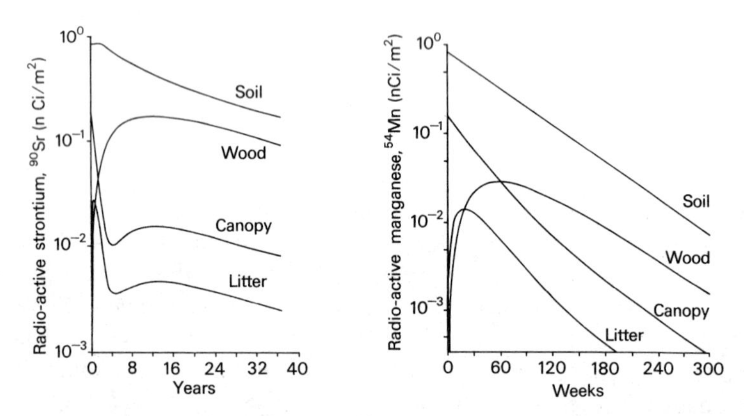

FIG. 6.7. Predicted storages of ^{54}Mn and ^{90}Sr in a Puerto Rican tropical rain forest using an instantaneous injection of 1 nCi/m^2 for each element. From C. F. Jordan *et al.* (1973). Reproduced from *Health Physics*, **24,** (1973), 477–89, by permission of the Health Physics Society.

has an earlier peak, 1964 in fact, owing to the relatively higher rate of physical decay of ^{54}Mn compared with ^{90}Sr; for the same reason, the secondary peaks due to recycling of ^{90}Sr do not appear in the case of ^{54}Mn. The model predicts that after 1966, the greatest amount of radioactivity will be in the soil, followed by wood, canopy, and litter; and concentration will be greatest in the canopy, followed by litter, soil, and wood. Unlike ^{90}Sr which tends to persist in the system, ^{54}Mn, because of its relatively rapid decay rate, is removed fairly fast.

The predictions using an instantaneous input of 1 nanoCurie per square metre (1 nCi/m^2) of ^{90}Sr and ^{54}Mn in the canopy and soil are shown in Figure 6.7. The diagrams emphasize the finding that ^{54}Mn is less persistent in the soil and other parts of the rain forest ecosystem than is ^{90}Sr, due to the more rapid radioactive decay of ^{54}Mn.

6.2.2. *Pesticides in ecosystems*

Another problem in ecosystem dynamics which can be partly tackled by compartment models is the long-term impact of pesticides and other chemical pollutants in ecosystems. O'Neill and Burke (1971), for instance, have developed a simple model of the movement of DDT and DDE in food chains leading to man. The model, summarized in Figure 6.8, was used to predict future concentrations of DDT and DDE in man under varying assumptions of DDT usage. Six cases were investigated and the results of each are listed in Table 6.5. As can be seen, if usage of DDT in the USA continues to be reduced at the present rate, or if reduced at twice the present rate, or if stopped altogether in 1972, the levels of DDT in human adipose tissue can be expected to decrease. Only if the application were increased by reverting to 1966 levels would the residual levels in humans increase.

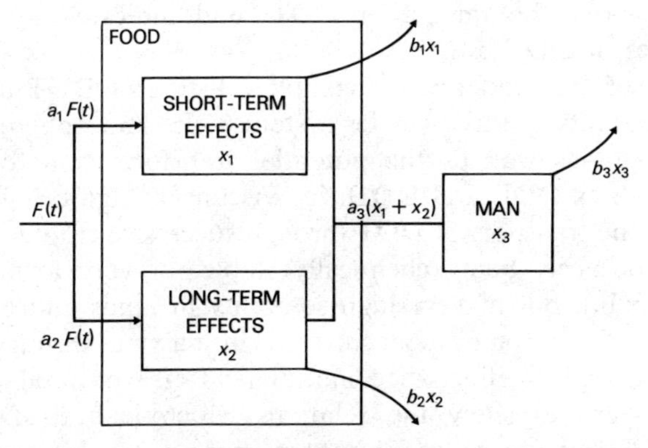

FIG. 6.8. DDT and DDE in food chains leading to man. Reprinted with permission, from R. V. O'Neill and O. W. Burke (1971).

Table 6.5
Predicted levels of DDT and DDE concentrations (p.p.m.) in human adipose tissue

Year	Case 1 Usage reduced at present rate	Case 2 Usage maintained at 1966 levels	Case 3 Usage reduced at twice present rate	Case 4 Maintained at 1972 level (5×10^6 lb /yr)	Case 5 Usage stopped in 1972	Case 6 Restrictions lifted in 1972, usage resumes at 1966 level
1970	5.14	5.30	4.98	5.14	5.14	5.14
1974	4.18	5.08	3.75	4.20	4.13	4.43
1978	3.41	5.32	3.00	3.53	3.35	4.82
1982	2.78	5.60	2.44	3.02	2.73	5.35
1986	2.27	5.84	1.99	2.60	2.23	5.82
1990	1.86	6.03	1.62	2.25	1.82	6.21
1994	1.52	6.20	1.33	1.98	1.49	6.52
1998	1.24	6.32	1.08	1.75	1.21	6.78
2002	1.01	6.43	0.88	1.56	0.99	7.00
2006	0.82	6.52	0.72	1.41	0.81	7.17
2010	0.67	6.59	0.58	1.29	0.66	7.31
2014	0.55	6.65	0.48	1.18	0.54	7.42
2018	0.45	6.70	0.39	1.10	0.44	7.52
2022	0.37	6.73	0.32	1.03	0.36	7.60

From O'Neill, R. V. and Burke, O. W. (1971). *A simple systems model for DDT and DDE movement in the human food-chain.* ORNL-IBP-71-9, Oak Ridge National Laboratory, Oak Ridge, Tennessee. By permission of the author.

Another interesting study of DDT dynamics is reported by Harrison *et al.* (1970), and involves the hearing conducted by the American State Department of Natural Resources considering a petition to ban the use of DDT in Wisconsin. The evidence included an integrated model of DDT transport which yielded testable hypotheses, demonstrated gaps in the knowledge about the impact of DDT, and indicated possible future consequences of DDT use.

The basis of the model was a complete listing of DDT inputs, transports, and outputs in three carrier systems—the atmosphere, water, and living biomass—as well as the potential transformation to breakdown products, such as DDE and DDD, in Wisconsin (Figure 6.9). Of special interest was the transport of DDT through successive trophic levels of the ecosystem, because, being chemically stable, of very low solubility in water (12 p.p.b.), but of very high solubility in lipids and other organic materials, DDT becomes concentrated in passing through successive trophic levels and its effects become inimical or even fatal. The aims of the model were to show the relations which lead to DDT and its metabolites being concentrated in the various trophic levels; and to indicate the dynamic nature of transport and concentrating processes.

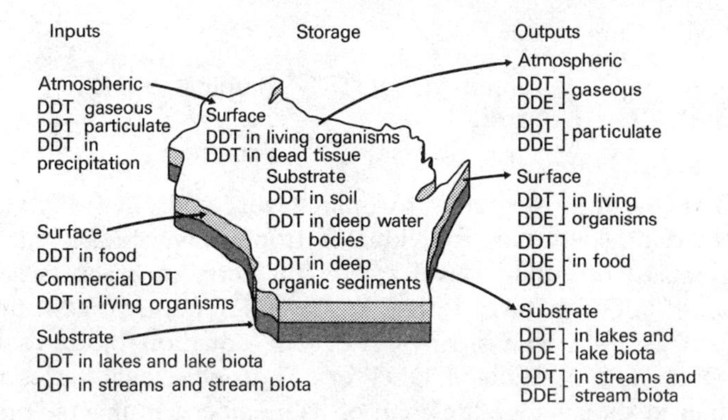

FIG. 6.9. DDT transport and storage in Wisconsin. From H. L. Harrison *et al.* (1970). Systems studies of DDT transport, *Science*, **170** (30 Oct. 1970), 503–8, Figure 1, with permission. Copyright 1970 by the American Association for the Advancement of Science.

DDT balance equations were used by Harrison *et al.* (1970) as a basis for a qualitative account of the pesticide transport through the Wisconsin ecosystem. Two interesting findings concern the steady-state concentration of DDT in different trophic levels, and the response time of the ecosystem—the time taken for a steady-state level of DDT to be reached.

Assuming that the Wisconsin ecosystem has reached a steady state with respect to DDT, that no subsequent addition of DDT is made to the system, and that metabolic breakdown is small compared with other terms (evidence for this assumption is available), then the steady-state concentration of DDT in any trophic level, C_i, has been shown by Harrison *et al.*, from a consideration of DDT balance equations, to be

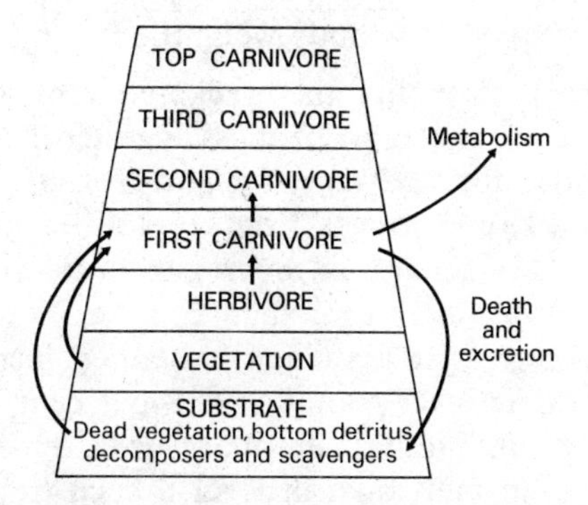

FIG. 6.10. Inputs and outputs of mass to the first carnivore level in the Wisconsin ecosystem. From H. L. Harrison *et al.* (1970), 'Systems studies of DDT transport', *Science*, **170** (30 Oct. 1970), 503–8, Figure 2, with permission. Copyright 1970 by the American Association for the Advancement of Science.

defined as follows

$$C_i = \frac{T_i}{x_i}\left(\begin{matrix}\text{inputs of DDT} \\ \text{from lower} \\ \text{trophic levels}\end{matrix} - \begin{matrix}\text{outputs of DDT} \\ \text{in} \\ \text{excretion}\end{matrix}\right)$$

where T_i is the turnover time of trophic level i and physically represents the average life-span of an individual in trophic level i; x_i is the standing crop (biomass) of trophic level i; and the term in brackets is the net retention of DDT in level i, that is, the difference between inputs and outputs of DDT in the steady state. The equation indicates that the steady-state concentration of DDT in any trophic level varies in direct proportion to the average life-span of its members, in inverse proportion to the biomass of the level, and in proportion to the net retention of DDT in the level. In moving through successively higher levels of the ecosystem average life-span increases and total biomass decreases; so, the equation implies that the concentration of DDT will increase in higher trophic levels. This finding is supported by a wealth of field evidence, such as the data assembled for Lake Michigan, where the concentration of DDT in bottom muds is 14 p.p.b., in amphipods 410 p.p.b., in fish 4500 p.p.b., and in herring gull at the top of the food chain, 99 000 p.p.b.

Assuming that organisms in all consumer levels feed only on organisms in the next lower trophic level, retain all DDT ingested, nor metabolize nor excrete DDT, Harrison *et al.* (1970) showed that the steady-state concentration of DDT in trophic level i, C_i, is proportional to the steady-state concentration of DDT in trophic level i-1, C_{i-1}, in the following way

$$C_i = \left(\frac{\text{mass ingested from trophic level } i-1 \text{ during life time}}{\text{body weight}}\right) C_{i-1}.$$

The coefficient of proportionality, the bracketed term, is, for an individual in the ith trophic level, the ratio of mass ingested from the next lower trophic level in a life time to body weight. Because this coefficient is always greater than 1, and ranges from 10 to 10 000, the steady-state concentrations of DDT increase in moving from lower to higher trophic levels for C_i must always be bigger than C_{i-1}.

It can also be shown that a sudden, sustained increase in the DDT concentration in trophic level i attains 98.2 per cent of its new steady-state value in a time equal to $4T_i$; in other words, remembering that T_i is average life-span of an individual in level i, each trophic level requires about four average life-spans of its members to reach steady state in response to a change in DDT concentration in the level below. Therefore, in ecosystems with long-lived members like the herring gull, with a life-span between 2.8 and 50 years, and the osprey, eagle, and falcon, with life-spans of as long as 60 to 100 years, the full effects of DDT (and many other pollutants for that matter) have not yet been felt nor will they be until later this century and into the next.

6.2.3. *Experimental component ecosystem models*

Some compartment models of ecosystems rely heavily on empirical relations to govern flows between compartments. Such models are called 'experimental component models' and are useful in making practical predictions about specific ecosystems. For instance, the management of upland nature reserves in Britain can be said to have three objectives; to maintain a vegetation cover sufficient to prevent soil erosion; to maintain the biotic, structural, and trophic diversity of varied ecosystems that are not already in that state; to preserve individual species of high interest, rarity, and vulnerability regardless of their functional position in an ecosystem (Milner, 1972). These objectives can usually be achieved by controlled grazing and burning and optimum results can be got from a management plan based on an experimental component model which will simulate system changes. A model constructed for the St. Kilda National Nature Reserve, as described by Milner (1972), will serve as a good example.

In the St. Kilda National Nature Reserve, an archipelago with four main islands and several isolated stacks, lies Hirta, the largest of the islands. Hirta is particularly amenable to being modelled: it is grazed by just one large and unmanaged herbivore, the Soay sheep, which has no predators; it has a limited range of structural vegetation types, being dominantly grassland and dwarf shrub heath (Figure 6.11); it being an island, important biological processes are restricted spatially. Thus the Hirta island ecosystem can be closed and its interactions reduced to a manageable number.

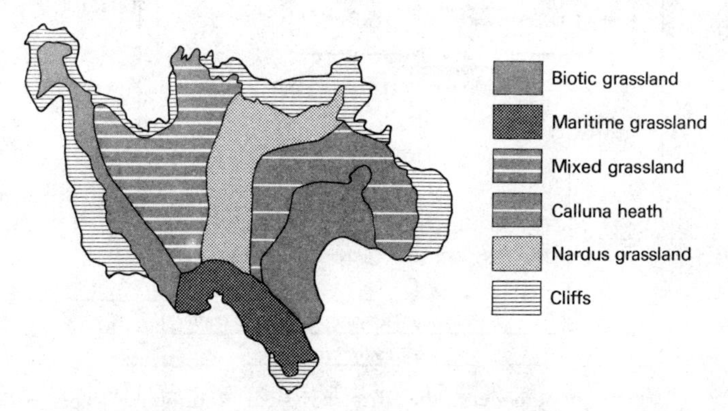

FIG. 6.11. Vegetation types on Hirta. Reprinted with permission from C. Milner (1972).

The system of interest for management purposes includes the sheep population and the vegetation on which they feed as the main components. The main compartments and flows of energy are depicted in Figure 6.12. In the model, use is made of realistic systems of differential and difference equations wherein exogenous variables, notably climate, and interactions between endogenous variables which are not capable of being expressed as a physical movement, such as competition for light or nutrients between different species in the same community, can affect the states and flows in the system. For example, plant growth and energy output by sheep are both considered weather dependent; and transfer rates are functions, not only of states in linked compartments, but also of compartments not directly linked, and of exogenous variables.

To operationalize the model it is translated into a computer language. The basic structure of the computer programme is shown as a flow chart in Figure 6.13. One run through the programme simulates one actual day. The main variables of the system are given appropriate initial values. The m communities, each of which consists of up to n plant species, of the

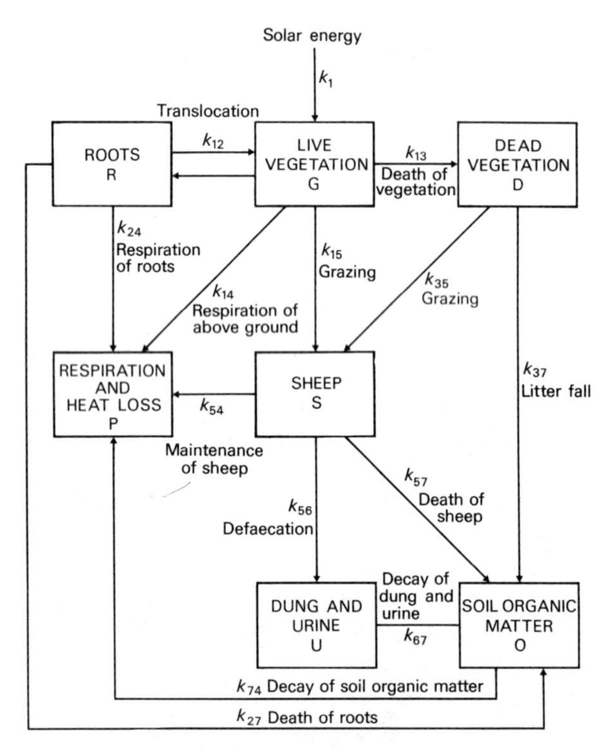

FIG. 6.12. A compartment model of the Hirta ecosystem. Reprinted with permission, from C. Milner (1972).

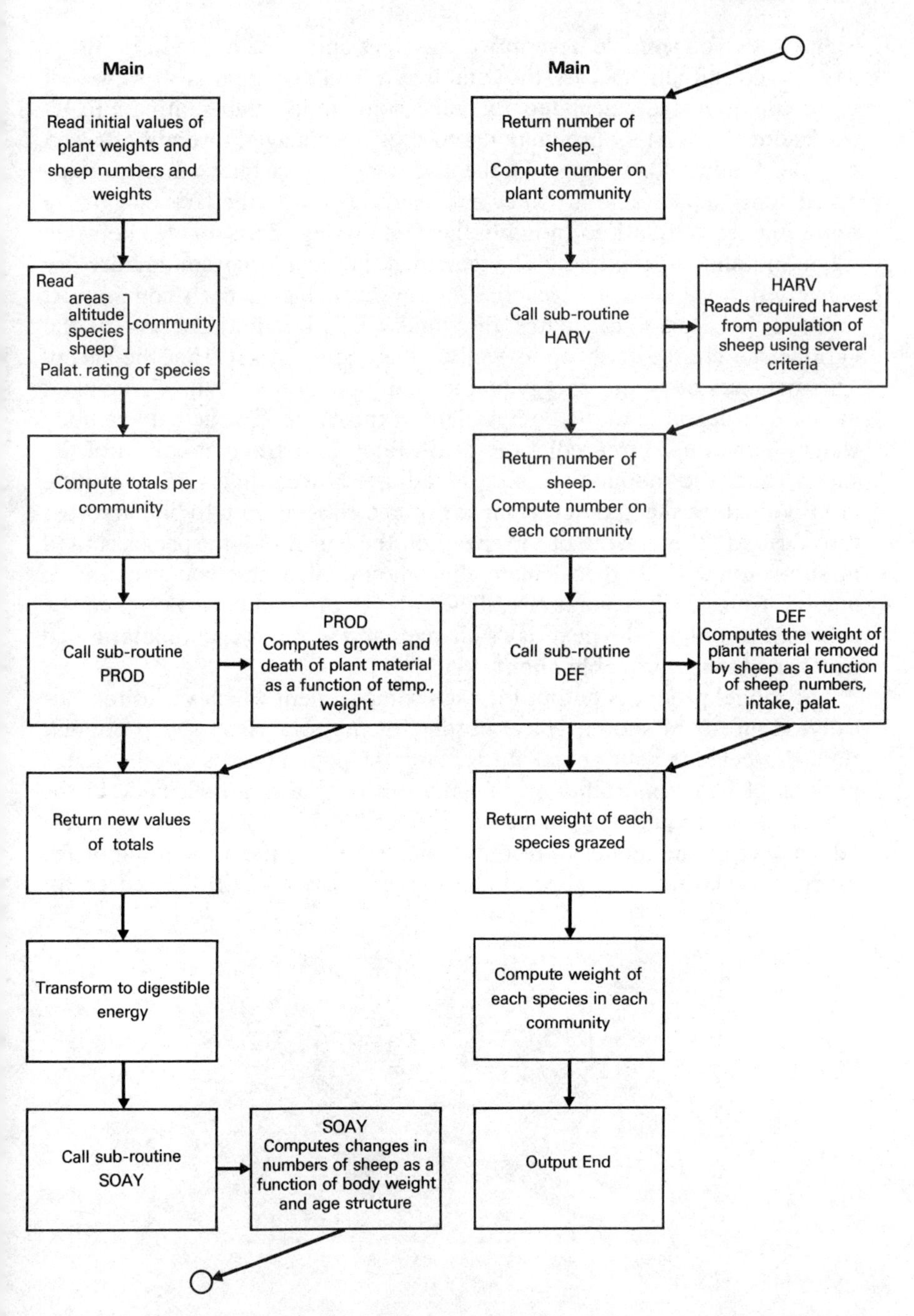

FIG. 6.13. Sturcture of the computer programme used to simulate the Hirta ecosystem. Reprinted with permission, from C. Milner (1972).

island are given suitable descriptive variables and these are read in. From data for constituent species, the total live weight and total dead weight of organic matter are calculated for each community. Sub-routine PROD works out the primary production and thus the change in weight of green and dead material. Change in the live-root compartment is also computed. This sub-routine is somewhat handicapped by the lack of data on some functional relations, notably the partitioning of assimilates between roots and shoots. Next, the changes in weight for each community are converted to figures of digestible energy available in each community. Sub-routine SOAY calculates the intake of digestible energy by the various age classes of sheep as well as the energy losses from the sheep. The balance between energy gains and energy losses may show an increase or decrease in the body weight of the sheep. The actual live body weight is then used to predict the death-rate of a particular cohort of the sheep, and the population size is adjusted accordingly. Sub-routine HARV enables sheep of a certain sex or age class or both to be harvested if so desired. The grazing of the sheep on the various plant species is used in sub-routine DEF to calculate the amount of each plant species removed from each community. The whole programme may then be repeated to predict the next day's changes and so on, for a predetermined number of runs, probably about 365.

The model produces output for each compartment which varies realistically. Figure 6.14 shows typical output for the total Soay sheep population. Agreement with actual fluctuations in population is good, though periods of heavy mortality are slightly out of phase, a deficiency in the model which could be corrected. Milner (1972) stated five concrete advantages of the model to resource managers: (1) the structuring of the model has clarified our view of the island ecosystem; (2) the search for

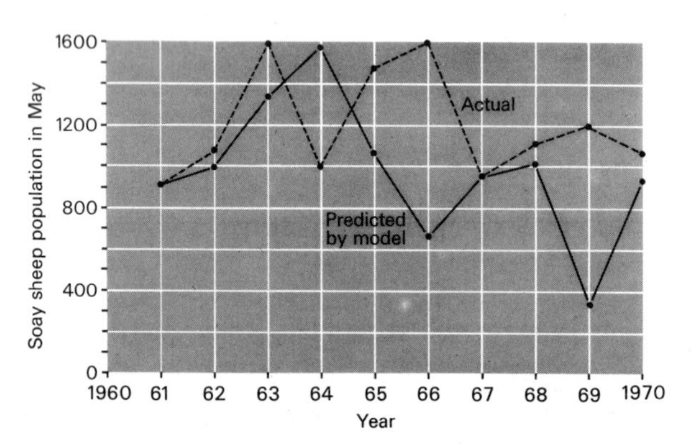

FIG. 6.14. Sample output from the Hirta simulation—changes in sheep population. Based on data in C. Milner (1972).

appropriate functions has generated much research which might not otherwise have been done; (3) the model has pinpointed those portions of a vast array of available data that are pertinent to study; and related to this, (4) its failings suggest directions for further field-data collections: for instance, if run for several years, the primary production and sheep population exhibit extreme fluctuations and this suggests that high priority should be given to research which examines this portion of the model; and (5) the model provides the manager of the reserve with a tool with which he can guide operations.

6.3. Hydrological models

6.3.1. *Catchment studies*

The best-known model of the land phase of the hydrological cycle is the Stanford IV Watershed Model. Devised by Crawford and Linsley (1966) and subsequently developed by others, Fleming (1970) and Wood and Sutherland (1970) for instance, this model aims to predict run-off from over-all drainage basis characteristics. In essence, the model divides a catchment into a number of water storage components. Water enters and leaves the system and moves from one store to another by the pathways shown in Figure 4.5. The rates of water transfer between stores are governed by parameters and functions which enable, for example, the physical characteristics to be altered and allowance made for varying soil moisture conditions within the catchment.

The Stanford Watershed Model has been used to predict the effects of urbanization on run-off characteristics. A similar kind of catchment component model was built by Wood (1974) specifically to study the changes in streamflow regime produced by urbanization. Wood's (1974) model is structured so that run-off could be separately predicted for the urban and rural portions of the catchment. Furthermore, it is designed so that, if desired, subcatchments can be studied; this is necessary for hourly input data of precipitation and for catchments larger than about 100 km^2. Five surface conditions are taken into account, two of which are urban and three rural (Figure 6.15). As in the Stanford model, the simulation is based on relations governing flows between the various catchment components, each of which can store water. In the case of urban run-off simulation, precipitation during a given time interval fills any space in infiltration storage, which is depleted by evaporation, and any excess is routed as overland flow storage, the water in which is fed into the sewer system. In the case of rural run-off simulation, run-off estimation centres on the calculation of soil moisture deficit, that is, the cumulative balance of inflow and outflow from soil-moisture storage, in all or any of three zones: a long-root zone of woodland; a short-root zone of grass and

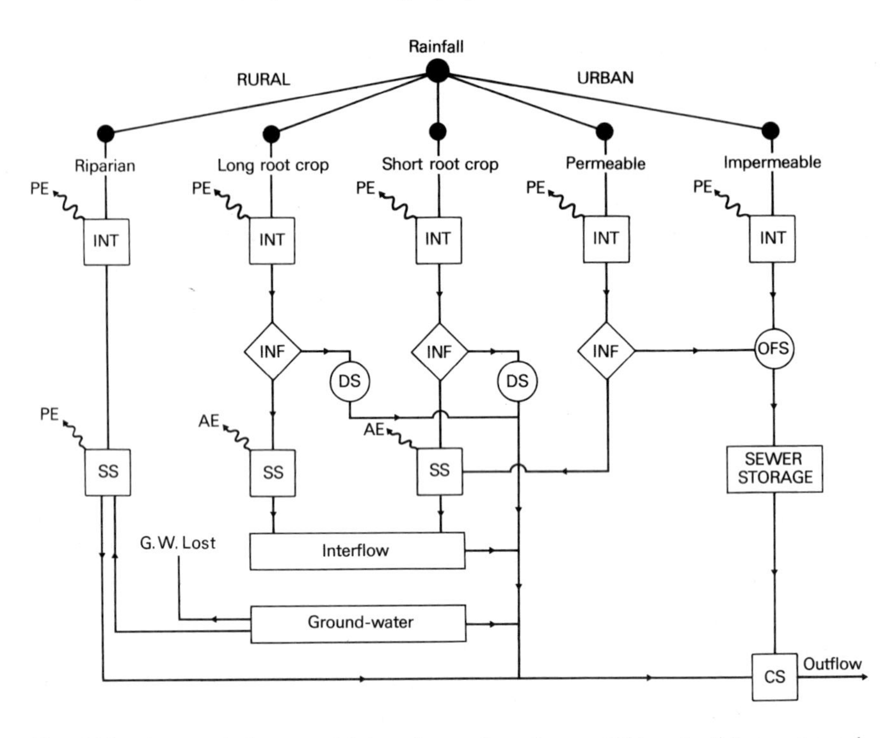

FIG. 6.15. A water-balance model for urban and rural cases. PE is potential evapotranspiration; AE is actual evapotranspiration; INT is interception storage, and INF is an infiltration function; DS is detention storage; OFS is overland flow storage; SS is soil storage; and CS is channel storage. Reprinted with permission, from S. R. Wood (1975). A catchment simulation model developed for urban and urbanizing catchments with particular reference to the use of automatic optimizing techniques. In *Modeling and simulation of water resource systems.* (ed. G. C. Vansteenkiste). North-Holland Publishing Company, Amsterdam.

various crops; and a riparian zone near watercourses where, owing to a near-surface water table, water supply is virtually unlimited. The soil-water model involves at each time step the balancing of soil-moisture deficit, actual evaporation, surface run-off, and seepage to other sub-surface stores such as interflow and ground water. Critical to how much evaporation occurs is the root constant which is simply the soil-moisture deficit that can accrue without evaporation being checked. Generally speaking, the outflow from all storages are defined by the following relation

$$Q_t = KS_t^a$$

where Q_t is the outflow from a store in a time interval t; S_t is the storage at the start of the time interval; K is a storage coefficient; and a is a

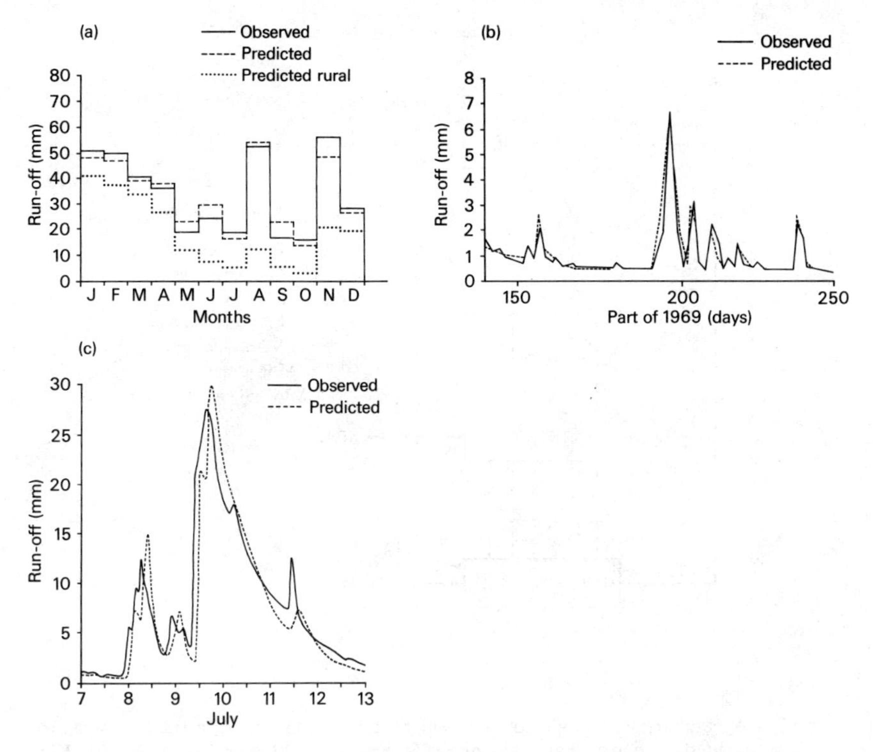

FIG. 6.16. Simulation output for the River Rea: (a) monthly yield for 1970; (b) daily flow for 1969; and (c) hourly flow for July 1968. Reprinted with permission, from S. R. Wood (1975), op. cit. Figure 6.15. above.

storage exponent. The model has been tested for two catchments in England—the River Rea catchment in the south-west suburbs of Birmingham and the Crawter's Brook catchment in Crawley, Sussex. Some results from the Rea catchment are shown in Figure 6.16; they show good agreement between predicted and observed flows for daily and hourly data and for monthly yield.

6.3.2. *Soil-water studies*

Coming down in scale from the catchment, various models have been constructed to similate the movement of water and the water balance in small-scale ecosystems. A good example is the model constructed by Sasscer *et al.* (1971) to study the movement of water through an old-field ecosystem in Argonne, Illinois. The system consists of one vegetation compartment and forty-nine soil compartments. Each soil compartment represents a one-centimetre-thick soil layer. The compartments and flows of water between them and links with the environment are portrayed in Figure 6.17. The behaviour of two types of water was modelled: stable

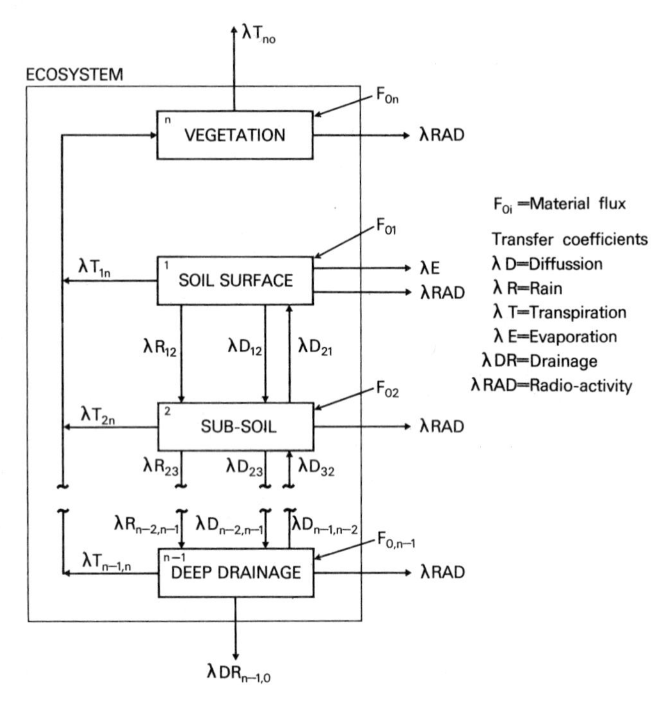

FIG. 6.17. A compartment model of material transfer paths through a soil profile. Reprinted with permission, from D. S. Sasscer, C. F. Jordan, and J. R. Kline (1971).

water and tritiated water (tritium). The fluxes or flow rates, all of which are considered to pass vertically through the compartments' surfaces, are measured in units of millilitres per square centimetre per hour (ml/cm^2/hr) for stable water movement, and in units of degenerations per minute per square centimetre per hour (dpm/cm^2/hr) for tritium movement. Flows originating outside the ecosystem are denoted by F_{oi} and flows between compartments by F_{ij}. The amount of water stored in a compartment per square centimetre of the surface area of the compartment at a given time is x_i. In general, the flux of water between compartments, F_{ij}, is proportional, by a turnover rate λ_{ij}, to the storage of the donor compartment:

$$F_{ij} = \lambda_{ij} x_i \quad (\text{ml/cm}^2/\text{hr or dpm/cm}^2/\text{hr}).$$

Sasscer *et al.* (1971) assumed the transfers between compartments are produced by diffusion, mass flow (owing to rain), and evaporation. Each of these transfer-mechanisms is incorporated into the model by giving a separate transfer coefficient for each: λD_{ij} for diffusion, λR_{ij} for mass flow, and λT_{ij} for evaporation. The general flux between compartments is thus given by

$$F_{ij} = (\lambda D_{ij} + \lambda R_{ij} + \lambda T_{ij})x_i$$

flux of water — diffusion rate — mass flow rate — evaporation rate — storage

Water may leave the ecosystem by evaporation from plants, λT_{n0}, evaporation from the soil surface, λE, and by deep drainage from the bottom-most compartment, $\lambda DR_{n\text{-}1,0}$. In the case of tritium, loss from each compartment occurs due to radio active decay, λRAD_{i0}. The storage of water in compartment j at time $t+1$, $x_{j,t+1}$, equals the storage of water in compartment j at time t plus the product of a state transition function, which takes the form of water inputs less water outputs, and the time interval:

$$x_{j,t+1} = x_{j,t} + \sum_{\substack{i=0 \\ i=j}}^{n} (\underbrace{\lambda_{ij}x_i}_{\text{inputs of water}} - \underbrace{\lambda_{ji}x_j}_{\text{outputs of water}})\Delta t, \qquad j = 1, 2, \ldots n.$$

With all loss and gain terms included (evaporation, diffusion, and so on) we have

$$x_{j,t+1} = x_{j,t} + \Bigg\{ \underbrace{\sum_{\substack{i=0 \\ i\neq j}}^{n} (\lambda D_{ij} + \lambda R_{ij} + \lambda T_{ij})x_i}_{\text{inputs of water}} - \underbrace{\sum_{\substack{i=0 \\ i\neq j}}^{n} (\lambda D_{ji} + \lambda R_{ji} + \lambda T_{ji} + \lambda E_{ji}}_{\text{outputs of water}} + \lambda DR_{n\text{-}1,0} + \lambda RAD_{j0})x_j \Bigg\} \Delta t.$$

The change of storage in each of the fifty compartments is thus represented by a first-order difference equation and the entire system is represented by a set of fifty simultaneous difference equations. This system of equations can be solved on a computer for given values of rainfall input, initial conditions, and transfer coefficients established by field experiment. Some results from the model are shown in Figure 6.18 which shows tritium concentration in a soil profile as a function of time. The two figures (6.18a and 6.18b) show the results of separate experiments, the first showing the pattern for 20 microCuries of tritiated water placed on the soil surface during a one-hour period; the second showing the pattern for 17.7 microCuries injected at a depth of 15 cm. Experimental data show very good agreement with the predicted results pointing to the validity of the model.

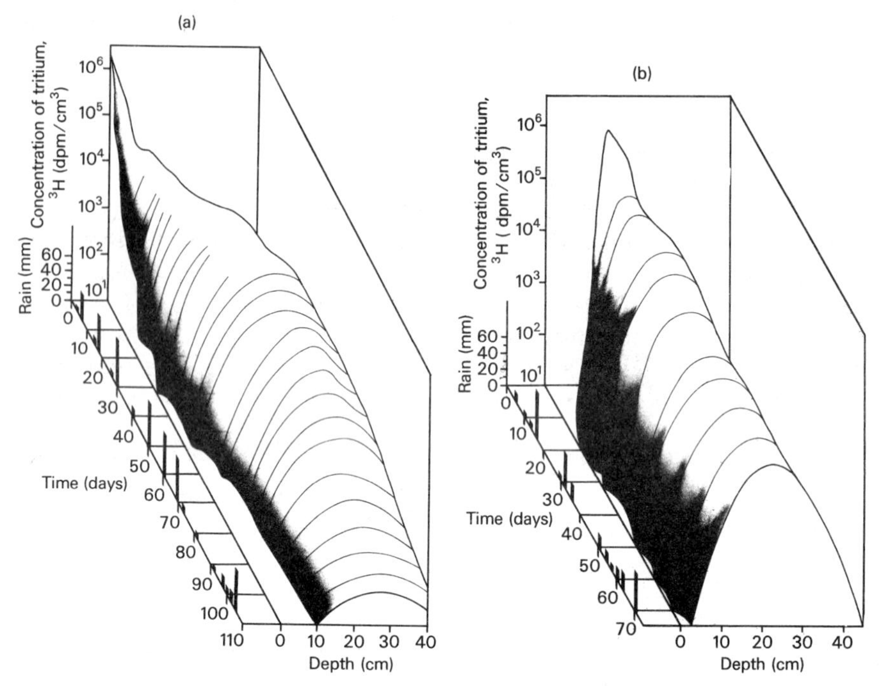

FIG. 6.18. Tritium concentration in soil as a function of time and depth for (a) deposition of tritium on the soil surface; and (b) deposition of tritium at 15 cm depth. Reprinted with permission, from D. S. Sasscer *et al.* (1971).

6.4. Socio-economic models

6.4.1. *A model of the United States*

Watt *et al.* (1975) designed a model to simulate certain changes in US society. Population, agricultural, land pricing, transport, and energy systems are included in the model, each of which is characterized by a number of descriptors and state variables (Figure 6.19). These variables are measured in the context of two hierarchical units: a composite urban area, CUA for short, which is a statistical average of 101 urbanized areas all surrounded by prime agricultural land (this accounts for 43 per cent of the US population); and a national level, the USA, which consists of two, interacting sectors—the crude-oil market and the agricultural market. The inclusion or exclusion of variables in this model was based on four criteria drawn from experience with statistical analysis of data and with computer simulation models of social systems. The first criterion was that only those variables and processes central to system functioning, in so far as they affect several other variables, influence long-term or medium-term trends, or are in any otherway critical to the system's behaviour,

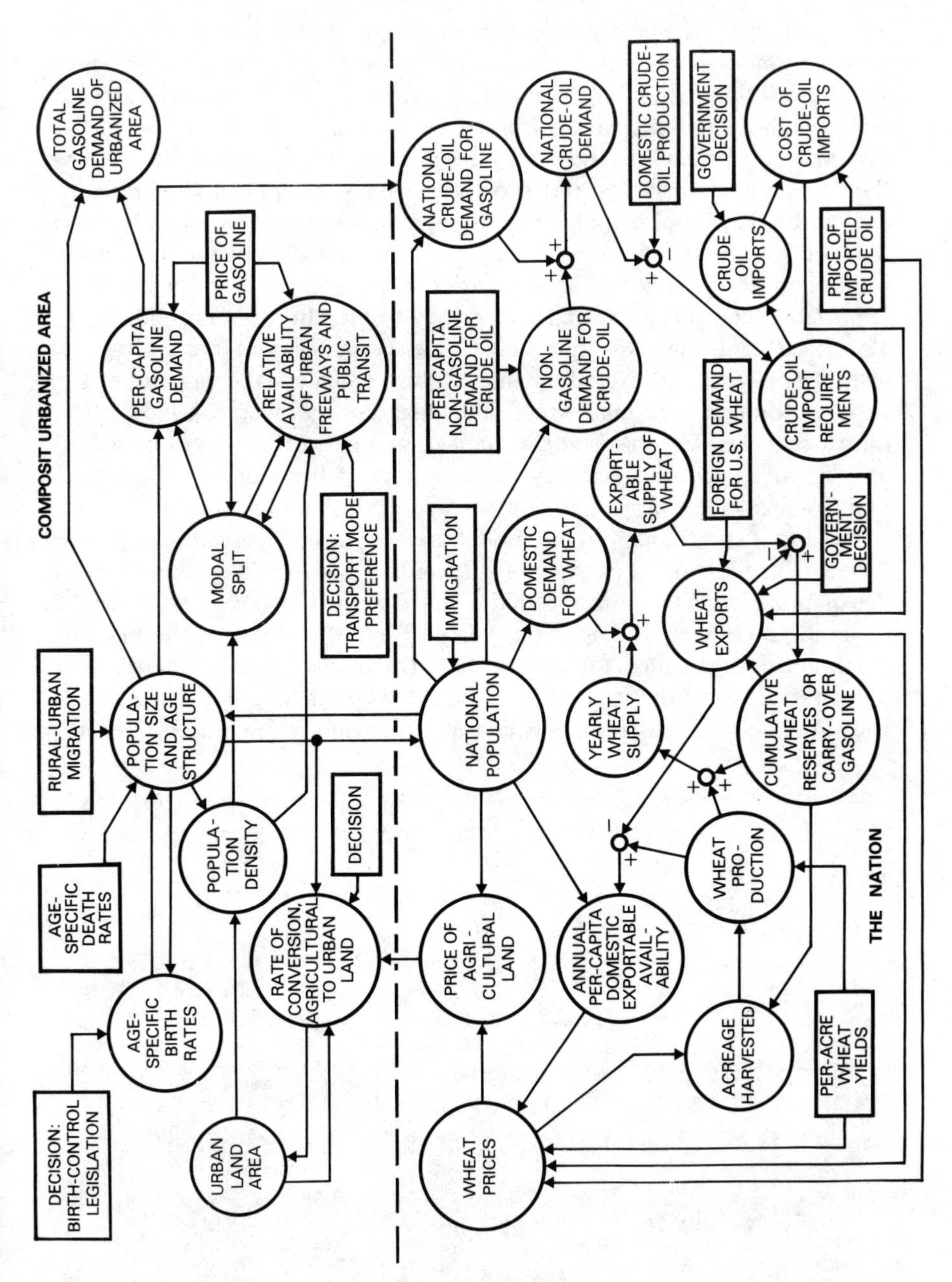

FIG. 6.19. State variables and relations for model to simulate certain features of US Society. Circles represent system variables; rectangles represent input variables. Reprinted with permission, from K. E. F. Watt *et al.* (1975). A simulation of the use of energy and land at the national level. *Simulation*, May 1975.

were included. Thus, investment in plant and equipment was excluded because under-investment and over-investment may merely produce short-term fluctuations—the business cycle—around a broader trend; demographic variables were included because they affect all else in the system, and particularly the demand variables. The second criterion was that only those variables and processes which change rapidly were included. Energy input from the sun, though an important variable, because it is constant, will not cause changes in society; social change may ensue from changes in rapidly changing variables such as total national cost of imported crude oil, price of food, price of agricultural land, and so on. The third criterion was that only those variables that are politically, institutionally, or individually managed were included: the vagaries of the weather, though they may wreak immense damage, were excluded. The fourth criterion was that some variables, though in themselves of secondary importance, are necessary in some part of the model; city population, for instance, was taken as a surrogate for traffic flow.

The input conditions for the model are shown in Figures 6.20a–f. The domestic production of petroleum (Figure 6.20a) is based on the National Petroleum Council's report called ***U.S. Energy Outlook.*** The price of crude oil, measured in dollars per barrel (Figure 6.20b), and the price of gasoline (Figure 6.20c), follow historical trends up to 1973 whence they stay constant until the year 2000. (The money units are actually constant 1967 dollars deflated using a wholesale price index for crude oil and a

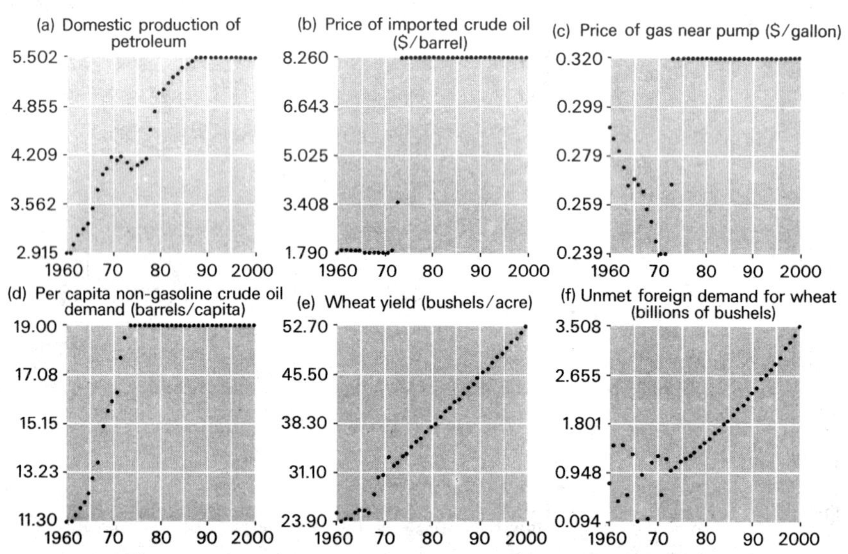

FIG. 6.20. Initial conditions. Reprinted with permission, from K. E. F. Watt *et al.* (1975), op. cit. Figure 6.19. above.

consumer price index for gasoline.) The per capita demand for crude oil in forms other than gasoline (Figure 6.20d) is assumed to remain constant after 1973 owing to the dampening effect of the high price of crude oil on demand. Wheat yield (Figure 6.20e) continues the historical increase from 1950. Foreign demand for US wheat (Figure 6.20f) is assumed to increase as shown on the figure which is based on world-wide estimates of supply and demand; this variable is in fact a source of considerable uncertainty and is strongly influenced by the weather.

The simulation output for some of the variables is shown in Figures 6.21a–i. Several findings are of interest. The conversion rate of prime agricultural land to urban uses decreases to 45 per cent of its 1970 value in the year 2000 (Figure 6.21a). Population density of the CUA decreases through 1975 but after 1985 increases, attaining a value 8 per cent above its 1974 level in the year 2000 (Figure 6.21b). The 1972 jump in gasoline

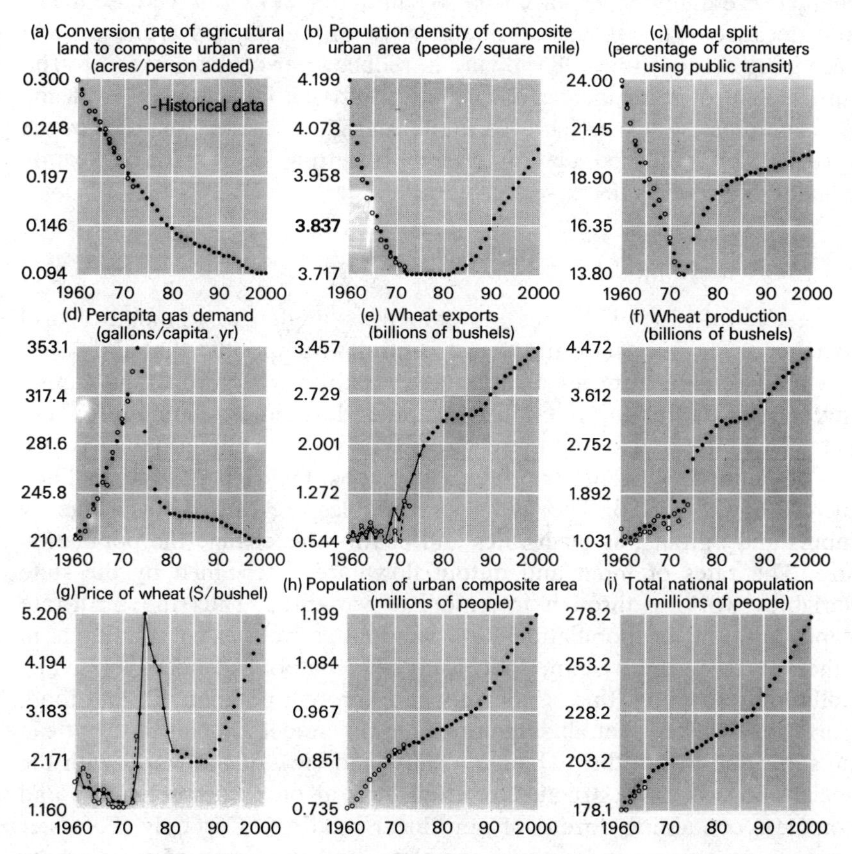

FIG. 6.21. Simulation output. Reprinted with permission, from K. E. F. Watt *et al.* (1975), op. cit. Figure 6.19. above.

steadily from 1972 (Figure 6.21c). The increase in use of public transit and the use of public transit (modal split): having decreased over the historical period from 1960 to 1972, the proportion of commuters using public transit and the ratio of public transit to freeway availability climbs steadily from 1972 (Figure 6.21c). The increase in use of public transit and increase in gasoline prices leads to a decreasing per capita demand for gasoline which drops to 70 per cent of its 1970 level in the year 2000 (Figure 6.21d) US wheat exports rise over the simulation period, and most rapidly between 1970 and 1980, mainly in response to large cost increases in imported crude oil (Figure 6.21e). Wheat production rises rapidly in the decade 1970 to 1980 in response to large exports (Figure 6.21f and g). The population growth in the CUA and the nation is shown in Figures 6.21h and i respectively.

The model output suggests three major response patterns which characterize many of the variables: (1) an initial and rapid increase in the first decade, owing to large price increases for imported crude oil and gasoline; (2) a drop to 1985, owing to reduced per capita demand for oil (in the light of price increases) and the additional impact of oil from the Alaskan pipeline; (3) an increase to the year 2000, caused for some of the variables by rapid population growth offsetting decreased per capita demand for crude oil.

6.4.2. *A world model*

Forrester (1971, 1972) has developed a model to simulate world dynamics. The model contains two central assumptions: that the world can be described in terms of five state variables, viz. population, capital investment, pollution, fraction of capital devoted to agriculture, and natural resources; and that these variables, with twelve intermediate ones, can be defined by twenty-two tables or graphs. Figure 6.22 is a simplified picture of the model. The five state variables are caused to change by inputs and outputs, as birth-rates and death-rates change the population size. The rates of input and output flows are determined by the state variables working through intermediate variables. Thus the birth-rate depends upon the population-state variable as well as on conditions in other parts of the system, specifically the material standard of living, pollution, crowding (this of course works through population), and food. Forrester believed that all assumptions in the model are plausible but can be subjected to scrutiny. Though a simplified view of the world with the social and economic structure omitted, it is at once comprehensive and complex, containing dozens of non-linear relations. Certainly, Forrester argued, the system is far more complete, and the result of running the model on a computer better, than the mental models at present in use for world and government planning.

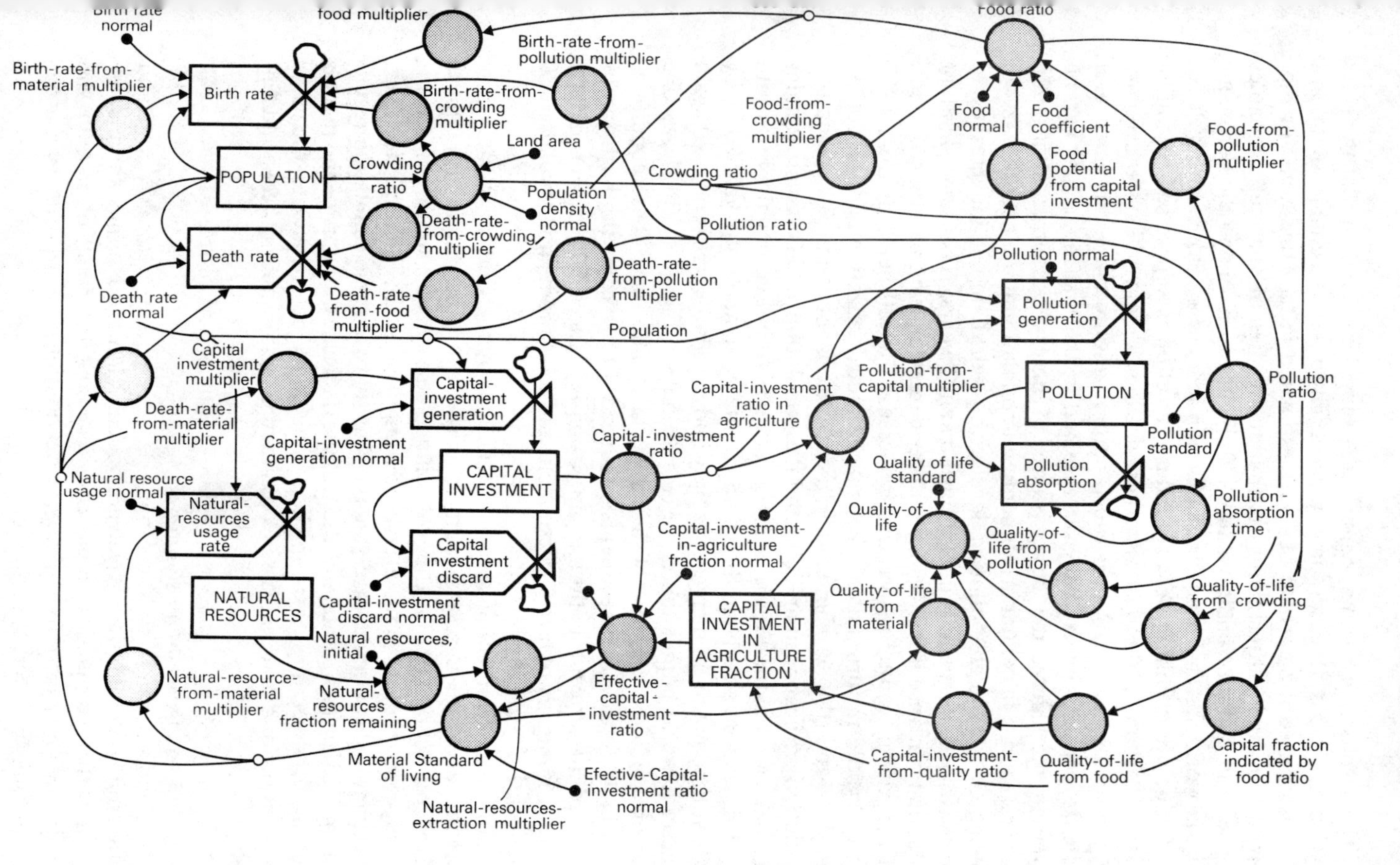

FIG. 6.22. **Structure of the world model Reprinted from *World dynamics* by J. W. Forrester by permission of the MIT Press, Cambridge, Massachusetts. Copyright © 1971 Wright–Allen Press, Inc.**

The results of a computer simulation show how the values of the state variables unfold through time from a given set of initial conditions. Forrester used a base date of 1900, estimated initial conditions for that time, and made adjustments so that the values of the state variables passed through the conditions of 1970. Figure 6.23a shows the behaviour of the world system assuming population reaches a peak at about 2020, subsequently declining because industrialization is suppressed by falling natural resources: the result is a slow decline in the world system. Figure 6.23b shows the situation in which natural resources are consumed at a rate 75 per cent lower than was assumed in the first model. As can be seen in the figure, the larger stocks of natural resources after 1975 generate a rise in capital investment and population until, about 2032–40, owing to very high pollution levels, the world population crashes to one-sixth of its peak value in just twenty years. The same kind of situation can arise if technology reduces dependence on natural resources. If the rate of capital investment is increased by 20 per cent from 1970 which would in turn increase the rate of industrialization and the quality of life—and this is the course the world is pursuing at present—a pollution crisis appears in 2030 and the population again crashes. Retaining decreased natural resource usage after 1970 but cutting the birth-rate by half leads to a surge in the quality of life up to about the year 2000, by which time pollution increases rapidly and a pollution crisis in 2020 again causes a world-population crash (Figure 6.23c). Figure 6.23d shows one set of conditions that establishes a world equilibrium. These conditions are: in 1970 capital investment is reduced 40 per cent, the birth-rate is reduced 30 per cent, pollution generation is reduced 50 per cent, natural resource usage is reduced 75 per cent, and food production is reduced 20 per cent.

Forrester's pioneering work was taken up by the Club of Rome whose preliminary investigations were published in *The Limits to Growth* (Meadows *et al.*, 1972). The world models have been attacked for using highly aggregated parameters, omitting important cultural feedback mechanisms, and under-stressing the sensitivity of some simulation results to small changes in certain of the state variables. Defenders of the models dismissed the critics as nit-pickers who could not invalidate the main conclusion that material-based economic growth cannot proceed indefinitely.

6.4.3. *A model of urban dynamics*

Forrester (1969) built a dynamic model of a city; it has three main components—businesses, jobs, and homes (see p. 66). The model was calibrated from numerous observed relationships between different levels

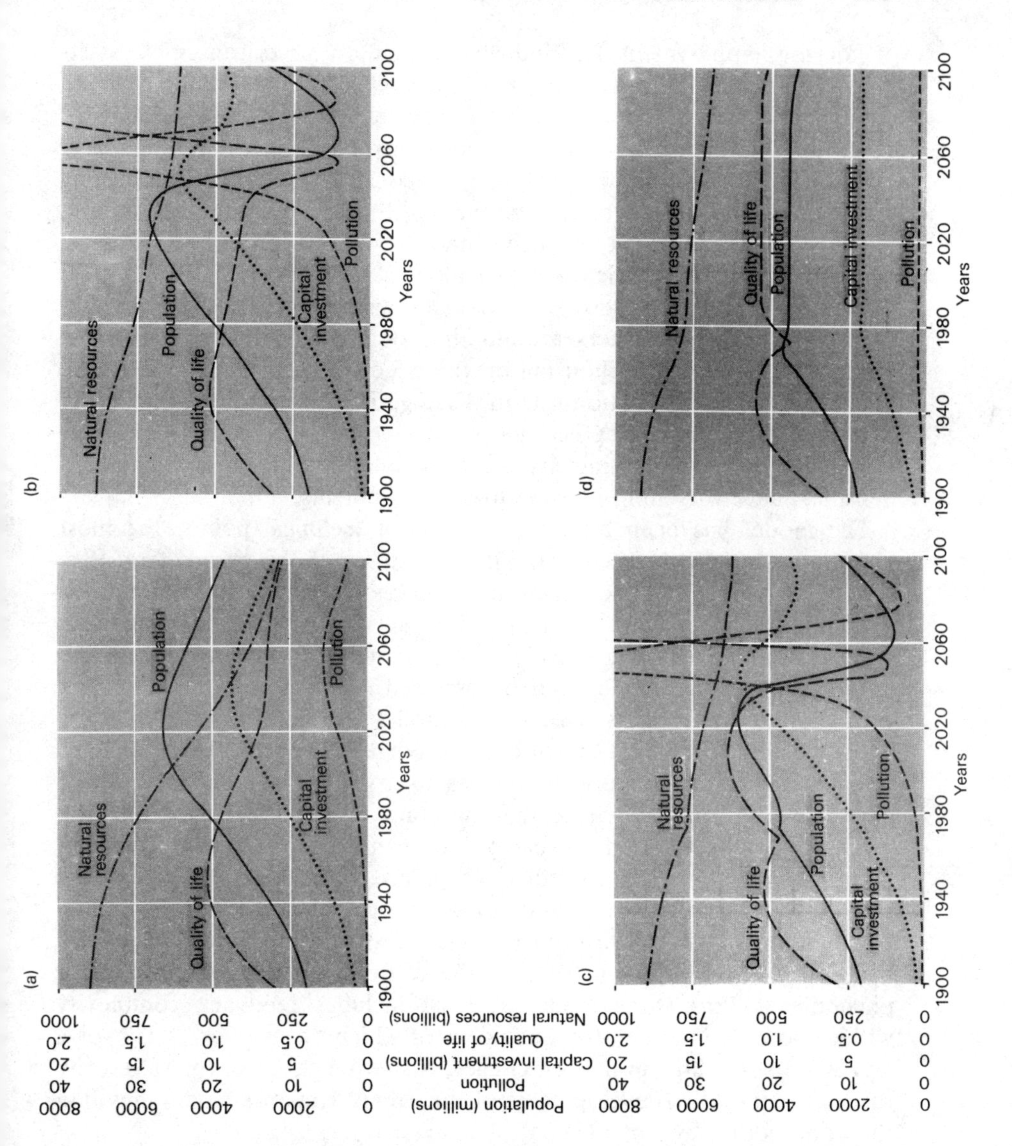

FIG. 6.23. Some simulation results. Reprinted from *World dynamics* by J. W. Forrester by permission of the MIT Press, Cambridge, Massachusetts. Copyright © 1971 Wright–Allen Press, Inc.

of housing, employment, and industry in modern American cities. Starting from an almost empty land area, the model predicted that a city will grow and in doing so pass through three stages. The first stage is a period of urban foundation and growth, which involves the establishment of businesses, the influx of workers, and the construction of housing, lasting a century in the model though, because of the nature of the calibration, the exact historical pace of change cannot be gauged. In the second stage, which runs for about eighty years, all the variables slump. In spatial terms, this reflects a run-down of original housing accommodating unskilled and lowerpaid workers around the commercial city core, and the building of new rings of housing on the edge of the original city. In the final stage, urban stagnation sets in. Change is slower than in the earlier two stages and, in geographical terms, produces a slow, outwards expansion of the city, a decreasing city-centre population, increasing population densities, and worsening journey-to-work problems.

The model has been criticized on several technical points, the most important among which concerns the sensitivity of the parameters. Forrester has argued that the system modelled is insensitive to all but a few parameter changes, and on this point he is supported by Chadwick (1971). In any event, the sensitivity of parameters, and any other of the assumptions in the model, can be assessed. Geographers are unhappy with the fixed city limits used in the model and certainly variable city limits would alter the pattern of change. Forrester has also been taken to task for advocating unpopular policies to avert the gloomy prospect of city stagnation; and it is true that sweeping conclusions based on little evidence, evidence based on scanty data, can be dangerous. But bear in mind the model does not predict the precise pattern of change but the general course of events likely to ensue from present-day relations among components in the American urban system. Most remedial urban policies have been drawn up from very simple cause-and-effect views of urban phenomena: Forrester's model at least includes feedback complexity within the urban system, the behaviour of which, to use Forrester's term and as we shall show in the last chaper, is counter-intuitive (p. 189). (The reader interested in developments of the urban dynamics model may find the book edited by Mass (1974) of interest.)

7 Regional Models

> Which some devout bunglers will undertake to manage and modelize.
>
> GAUDEN, *Tears of the Church*

THE integration of system structure and function, of form and process, is seen in regional systems. In this chapter, we shall deal with the modelling and analysis of regional systems, systems in which the spatial location of flows and their connection with system structure are of paramount interest. Two types of regional system models may be recognized. Type one is those models in which the mutual adjustment of system form and system process is not considered but in which the spatial location of system processes is a prime factor. In the second type of regional-system model, the mutual adjustment of system form and system process is the chief concern.

Of the first type of regional models, the following will be considered: inter-regional compartment models, inter-regional population models, spatial-interaction models, and stochastic models. The second type of regional model will be exemplified by process–response models, both physical and human.

7.1. Inter-regional compartment models

The difference between the compartment models dealt with in chapter 6 and those that will be discussed here is that the former are not concerned with the spatial location of flows between compartments, whereas the latter are. In an inter-regional or spatial compartment model, a system is represented not just as a series of storage compartments but as a set of storage compartments that have a distinct shape, area, or volume, and a geographical location and boundaries.

7.1.1. *The geometry of spatial compartments*

The spatial compartments, or segments as they are usually called, may be arbitrary in shape or size, their geometry being varied according to the degree of detail required and the dimensions of interest. Thus, a study of eutrophication in western Lake Erie, which will be described in detail in the next subsection, used seven spatial segments which did not take account of the vertical changes in the lake water, this dimension entering the model in the volume of each segment. In studies where vertical stratification of temperature and density in lakes is pronounced, horizontal slices are employed as segments (Figure 7.1). Other situations require a different geometrical representation. Chen and Orlob (1975) represent estuarine geometry by dividing the water body into discrete volume units or nodes, each of which has surface area, depth, volume, and side-slopes. Nodes are connected by channels or links, each link being characterized by length, width, cross-sectional area, hydraulic radius (depth), and a friction factor (Figure 7.2). Water moves from node to node along the links and in doing so carries water-quality constituents which diffuse in the process. The link–node network can represent a range of estuarine conditions including a system of discrete estuarine channels and a combination of channels and shallow bays (Figure 7.3). This kind of spatial representation has been found capable of reproducing with good fidelity the hydrodynamic behaviour of estuarine water.

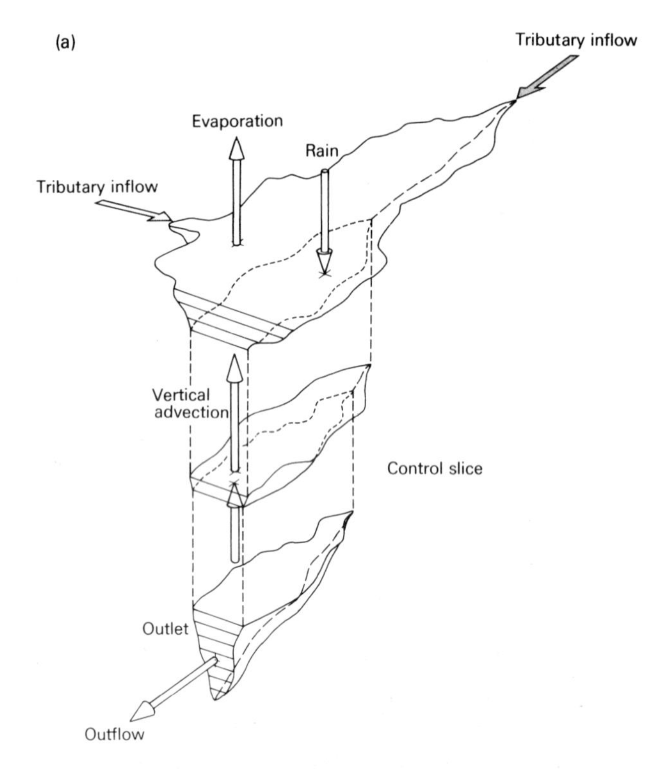

FIG. 7.1. (a) A geometrical representation of a stratified lake. (b) Water movement along three adjacent lake elements. Reprinted with permission, from C. W. Chen and G. T. Orlob (1975), 'Ecologic simulation for aquatic environments', in *Systems analysis and simulation in ecology*, vol. 3 (ed. B. C. Patten), pp. 475–588, Academic Press, New York.

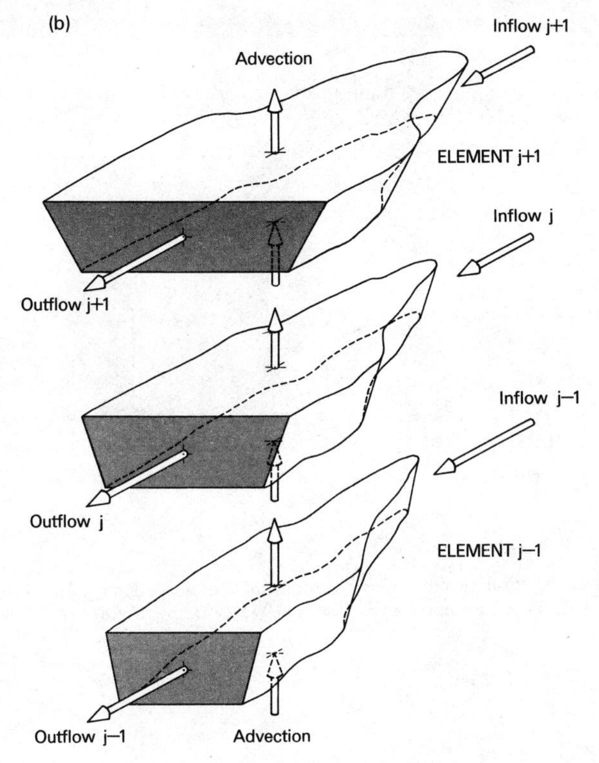

FIG. 7.1. (b)

Simpler spatial forms are sometimes used. Briggs and Pollack (1967) used a series of square cells to represent an ancient evaporite basin, the pattern of water flow, salt movement, and salt deposition in which they modelled with great success. Huggett (1975) used a block composed of cubes to represent a soil landscape system and was able to model the general, three-dimensional translocation pattern of mobile soil constituents in a landscape.

7.1.2. *Eutrophication in Lake Erie*

To see how spatial-compartment models are built and analysed, it is best to take a concrete example. We shall consider the phytoplankton-zooplankton-nutrient model for western Lake Erie as described by Di Toro *et al.* (1975); this model was constructed partly from pure scientific interest and partly as an aid to the understanding, management, and control of eutrophication in natural waters.

The model consisted of seven endogenous-state variables (Figure 7.4). A circulation of nutrients through the system is apparent: primary production converts inorganic nutrients into the biomass of phytoplankton;

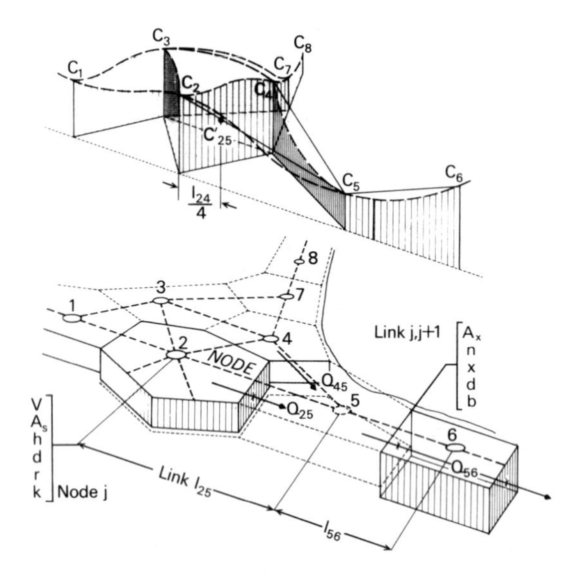

FIG. 7.2. Typical section of an estuarine node-link network. Reprinted with permission, from C. W. Chen and G. T. Orlob (1975), op. cit. Figure 7.1 above.

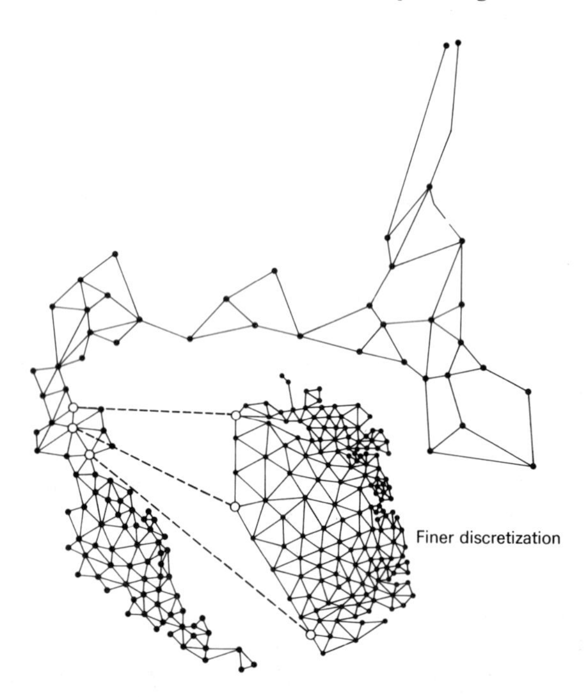

FIG. 7.3. Example network for the San Francisco Bay-Delta system. Reprinted with permission, from C. W. Chen and G. T. Orlob (1975), op. cit. Figure 7.1. above.

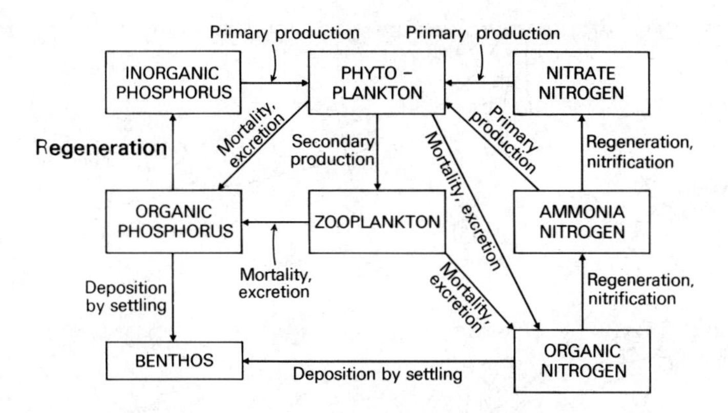

FIG. 7.4. A compartment model of the Lake Erie ecosystem. Reprinted with permission, from D. M. Di Toro *et al.* (1975), 'Phytoplankton-zooplankton nutrient interaction model for western Lake Erie', In *Systems analysis and simulation in ecology*, vol. 3 (ed. B. C. Patten), pp. 423–74, Academic Press, New York.

zooplankton graze on the phytoplankton and incorporate minerals in their bodies—secondary production; the demise of organisms and excretion release detrital and soluble forms of the nutrients; regeneration processes convert organic forms of the nutrients to inorganic forms which may be re-used by the phytoplankton; deposition of organic matter may occur. Exogenous variables considered in the model are physical ones—temperature, solar radiation, photoperiod, lake level, inflow from the Detroit river and water clarity; chemical ones—the chemical quality of the Detroit river and of the Maumee river; and biological variables—the phytomass and zoomass of the Detroit and Maumee rivers.

The section of Lake Erie that was modelled was divided into seven spatial segments (Figure 7.5). The basis of the model was a set of mass balance equations, one for each endogenous variable, which took into account flows between endogenous variables, flows between endogenous and exogenous variables, and flows between adjacent spatial segments. With seven state variables and seven spatial segments, a total of forty-nine compartments was studied. Flow between spatial segments was assumed to be the result of both advection—the bulk movement of water by lake currents, and dispersion—the diffusion spread caused by smaller water circulations which bring about mixing of lake water. Advection and dispersion terms are shown in Figure 7.6. The mass-balance equations took the general form:

$$V_j c_{i,j,t+1} = V_j c_{i,j,t} + \left\{ \sum_k Q_{k,j} c_{i,k} + \sum E'_{k,j}(c_{i,k} - c_{i,j}) + \sum_k S_{i,j,k} \right\} \Delta t$$

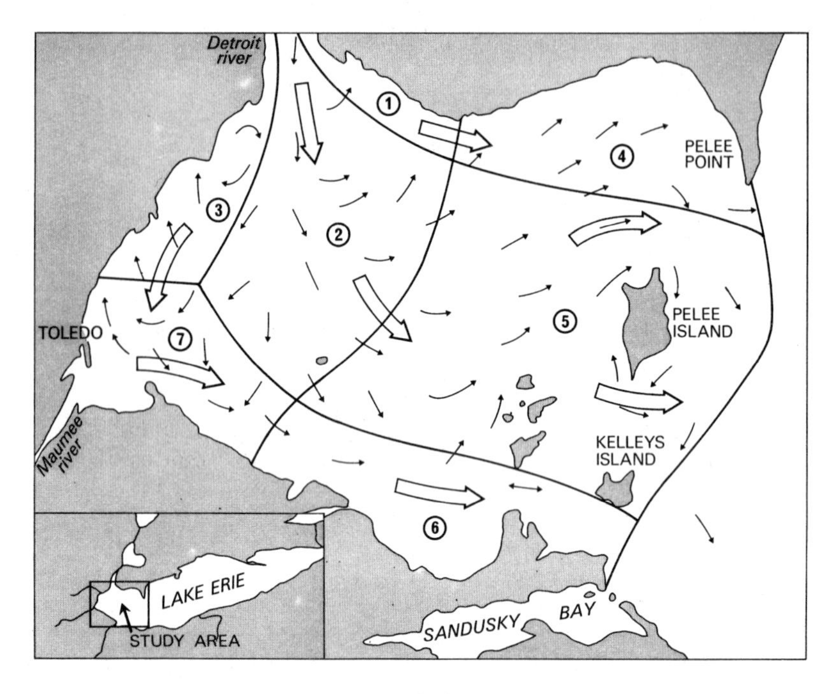

FIG. 7.5. Spatial segments used in the Lake Erie model. The prevailing currents directions are shown. Reprinted with permission, from D. M. Di Toro *et al.* (1975), op. cit Figure 7.4 above.

where V_j is the volume of spatial segment j; $S_{i,j,k}$ is the kth source (+) or sink (−) of substance i in segment j; $E'_{k,j}$ is the bulk rate of transport of $c_{i,k}$ into and $c_{i,j}$ out of segment j for all segments, k, adjacent to segment j; and $Q_{k,j}$ is the net advective flow rate between segments k and j. The equation for the phytoplankton is a balance equation for phytoplankton biomass, measured as chlorophyll-a concentrations, which equates rate of phytomass change to transport of phytomass by advection from adjacent segments, $\sum_k Q_{k,j}P_k$, and by dispersion into, $E'_{k,j}P_k$, and out of $-E'_{k,j}P_j$, segment j, as well as the growth rate, $G_{pj}P_jV_j$, and the death rate, $D_{pj}P_jV_j$, in segment j:

$$V_jP_{j,t+1} = V_jP_{j,t} + \left\{ \sum_k k_jP_k + \sum E'_{k,j}P_k - \sum E'_{k,j}P_j + G_{pj}P_jV_j - D_{pj}P_jV_j \right\} \Delta t.$$

phytomass in segment j at time $t+1$	phytomass in segment j at time t	net advection	dispersion into segment j	dispersion out of segment j	growth rate of phytomass	death rate of phytomass

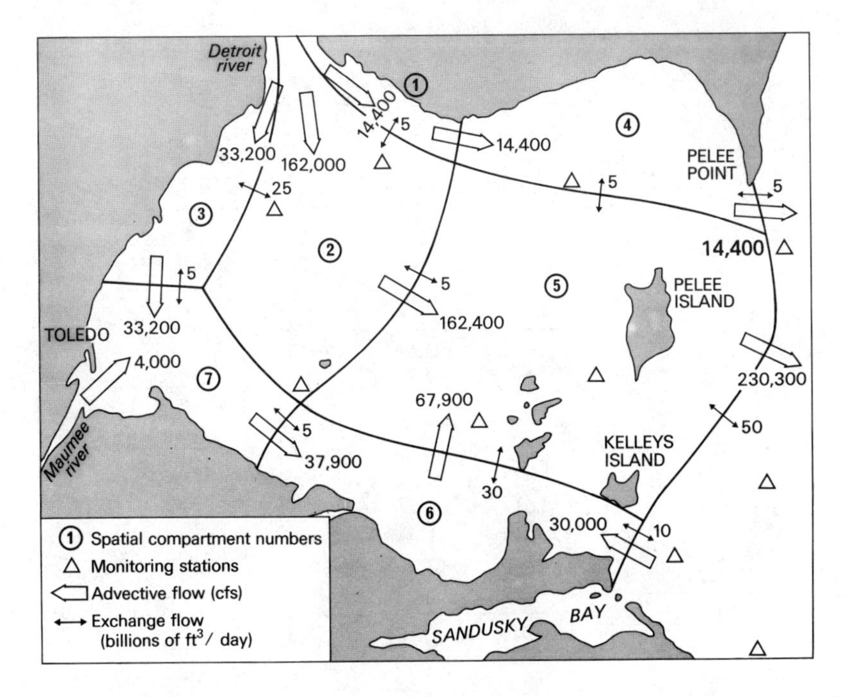

FIG. 7.6. Steady-state water transport for seven-compartment western Lake Erie model. Reprinted with permission, from D. M. Di Toro *et al.* (1975), op. cit. Figure 7.4 above.

A similar equation was developed for the zoomass in each segment. Boundary conditions were specified for the tributary rivers and the open-water boundaries of segments 4, 5, and 6. The parameters of the full set of differential equations were established from laboratory and experimental data.

The results consisted of predictions of nutrient concentrations made for each of the seven state variables in each of the seven spatial segments for the period April to November. These predictions were compared with actual observations made in the lake in 1968 and 1970; agreement between observations and predictions was quite good, though some deviations were found. Figure 7.7 shows the nitrate–nitrogen predictions, measured in milligrammes of nitrogen per litre; the major features of the data are well reproduced by the model. The model was also used to assess the effects on eutrophication of management alternatives. The eutrophication of the western part of Lake Erie has increased over the past fifty years. To predict future trends, nutrient inputs to the western end of the lake were computed on the basis of three different population projections

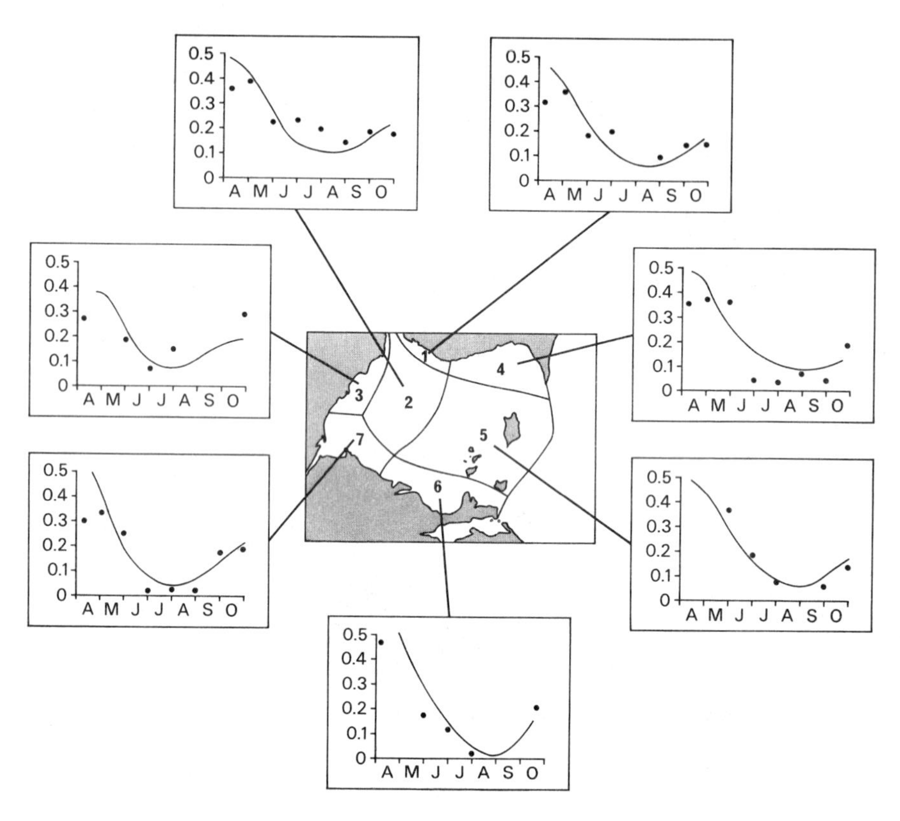

FIG. 7.7. Nitrate nitrogen in seven segments of western Lake Erie: observed and predicted data. Units are milligrammes of nitrogen per litre (mg N/l). Reprinted with permission, from D. M. Di Toro *et al.* (1975), op. cit Figure 7.4 above.

for the Lake Erie basin, the increase of population leading to an increase in urban run-off and industrial contributions. The results are shown in Table 7.1 for phytomass (summer average) in segment 7, adjacent to the Maumee river.

A second set of predictions was made to investigate the efficacy of removing phosphorous entering the basin. The results of an 80 per cent removal policy and a 95 per cent removal policy coupled with a total ban on detergent phosphorus using a moderate population projection are shown in Table 7.1. A third prediction tested the policy of 80 per cent phosphorous removal and 50 per cent nitrogen removal in water being discharged directly into the basin; this policy, it would seem, is capable of returning the lake to its 1930 condition by the year 2010.

Table 7.1
Eutrophication predictions for western Erie basin

Year	Observed phytomass (μg/l)	Population: accelerated growth	moderate growth	limited growth	Removal strategy: 1	2	3
1930	15	—	—	—	—	—	—
1970	25	—	—	—	—	—	—
1990	—	37	30	—	15	12	11
2010	—	42	35	26	20	22	16

Removal strategy: 1. 80 per cent removal of phosphorous; 2. 95 per cent removal of phosphorous and detergent; 3. 80 per cent phosphorous and 50 per cent nitrogen removal.
Source: based on Di Toro *et al.* (1975).

7.1.3. *Compartment models for comparative purposes*

Some compartment models, without being overtly spatial in having spatial segments, do consider locational differences in system behaviour. By examining system dynamics using a compartment model and making several simulations, the parameters in each being applicable to different environments, insight can be gained into the reasons why say nutrient cycles vary in different parts of the globe. Gersmehl (1976) proposed a general, three-compartment model of mineral cycle (Figure 7.8) and showed how this model can shed light on the differences in mineral flow and storage patterns in the world's biomes. The nutrient circulations of some major terrestrial ecosystems, each represented as a three-compartment model, are shown in Figure 7.9. Gersmehl (1976) argued

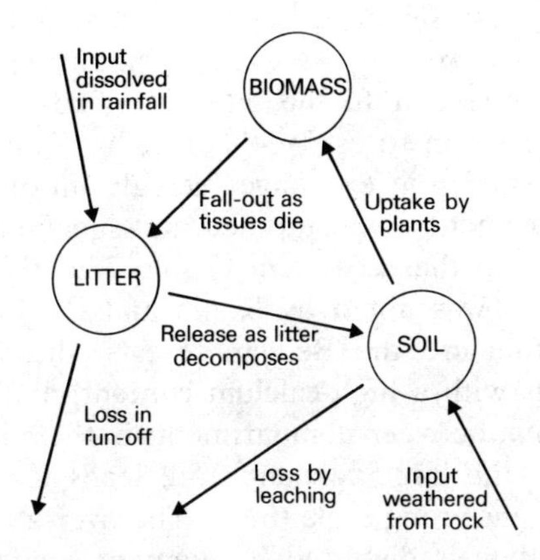

FIG. 7.8. A general, three-compartment model of a mineral cycle. Reproduced by permission, from the *Annals* of the Association of American Geographers, **66** (1976), P. Gersmehl.

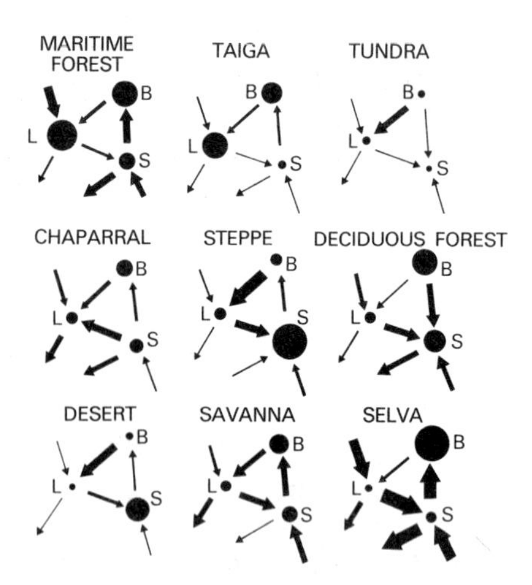

FIG. 7.9. The nutrient circulation in some of the world's major biomes. Reproduced by permission, from the *Annals* of the Association of American Geographers, Volume 66, 1976, P. J. Gersmehl.

that the main differences between circulations in ecosystems result from differences in driving functions—temperature, which reflects broad differences in surface-energy budgets, and precipitation; differences in driving functions are large enough to impart to ecosystems regionally unique patterns of mineral circulation. (The interested reader should consult Gersmehl's article for a description and explanation of the major circulations.)

In a similar vein, Jordan and Kline (1972) sought to establish a world pattern of mineral cycles using a compartment model of the calcium cycle into which were fed data from ecosystems ranging from the tropics to the taiga. The model was similar to the one described for manganese and strontium in a tropical rain forest (p. 112); it has four compartments—wood, canopy, litter, and soil (exchangeable calcium only). The turnover time for each compartment was computed and summed to give the total cycle time of calcium in the ecosystem (Table 7.2). Cycle times tend to increase polewards, as was noted by Rodin and Basilevich (1967). The exceptions to this rule are the Belgian forests; these forests grow in calcareous soils (soils with a high calcium content) and so, although the actual fluxes of calcium between compartments in them is similar to fluxes in other ecosystems, the large calcium store leads to their having long turnover times and hence long cycle times. The over-all relation between cycle times and latitude is distorted by element availability in the soil compartment which is unrelated to latitude: by subtracting the turnover time in the soil compartment from the cycle time the pattern is improved.

Table 7.2
Compartment turnover times and cycling times for calcium

1	2	3				4	5	6	7
		Turnover time (yr)							
Ecosystem	*Type of ecosystem*	*Soil*	*Wood*	*Canopy*	*Litter*	*Total cycle time (yr)*	*Total minus soil (yr)*	*Latitude (°N)*	*Length of growing season (days)*
Puerto Rico	montane tropical rain forest	3.0	6.4	0.9	0.2	10.5	7.5	18	365
Ghana	moist tropical forest	8.2	6.8	1.5	0.2	16.7	8.5	7	240
England	Scots pine plantation	11.2	6.1	0.8	3.4	21.5	10.3	53	200
Virelles, Belgium	mixed oak	108.8	12.6	0.4	0.9	122.7	13.9	51	200
Wéve, Belgium	oak-ash	184.9	21.5	0.4	0.6	207.4	22.5	51	200
Washington State, USA	Douglas-fir	57.4	20.2	5.5	10.2	93.3	35.9	48	200
Southern taiga, Russia	spruce	5.1	22.2	1.3	9.0	37.6	32.5	55	120
New Hampshire, USA	northern hardwood	14.0	10.8	0.8	34.8	60.4	46.4	44	120
Central taiga, Russia	spruce	7.6	18.3	3.2	13.6	42.7	35.1	65	105

From Jordan, C. F. and Kline, J. R. (1972), 'Mineral cycling: some basic concepts and their application in a tropical rain forest', *Annual Review of Ecology and Systematics*, **3,** 33–50. Reproduced with permission.

If length of growing season is substituted for latitude an even better correspondence is obtained (Table 7.2). Jordan and Kline (1972) explained the global-cycle time pattern in terms of biomass production which, on a world scale and assuming that nutrient availability in soils is not exceptionally low or high, is highest in the tropics, decreasing with amount of solar radiation available during the growing season and increasing length of dry season.

7.2. Inter-regional population models

Models available for analysing and forecasting population change fall into two broad groups: components-of-change models, which deal with total births, deaths, and migrations; and cohort-survival models, which consider the behaviour of population cohorts, usually of 1000 individuals, disaggregated by age or sex or both (Rogers, 1968). The first built components-of-change models may be attributed to L. A. J. Quetelet, a Belgian astronomer of the nineteenth century, who considered population growth in a single region and, under assumptions of uninhibited and inhibited (resource-limited) increase, arrived at the patterns of exponential and logistic growth (see pp. 85–6). Cohort-survival models were first built by Leslie (1945) and, independently, by Lewis (1942). Leslie's basic model, though dividing a population into age and sex classes, was originally formulated for the population in a single region and, in fact, depending on the assumptions made, tends to describe either exponential or logistic growth of a cohort with a stable age–sex structure. A third group of population models considers competition and predation between two or more populations. Again, these models, as originally formulated, concerned populations in single regions.

In this book we are interested in the extension of the three basic types of population model to inter-regional cases. Nevertheless, to comprehend their inter-regional versions, an outline of the original models must be given.

7.2.1. *Components-of-change models*

In these models, population size (or density), N, is a state variable. It is assumed that the population at time $t+1$ is equal to the population one time step earlier, that is at time t, plus the net change in the population, brought about by births and immigration, by deaths and emigration, during the time period. Symbolically,

$$N_{t+1} = N_t + (\text{births} + \text{immigration} - \text{deaths} - \text{emigration})\,\Delta t. \quad (7.1)$$

The components-of-change terms—births, deaths, and migration—form the state transition function and are defined as proportional, by specific birth, death, and migration rates (denoted by the letters b, d, and m respectively), to the population size at time t; so we may write

$$\text{births} = bN_t$$
$$\text{deaths} = dN_t$$
$$\text{migration} = mN_t$$

where b, d, and m are analogous to rate constants in compartment models and would be measured in units of births, or deaths, or migrations per unit of population (usually per person or per 1000 people) per unit of time. For a system in which migration is absent, equation 7.1 could be written

$$N_{t+1} = N_t + (bN_t - dN_t)\,\Delta t$$

or

$$N_{t+1} = N_t + (b-d)N_t\,\Delta t. \tag{7.2}$$

The parameter $(b-d)$, the specific birth-rate less the specific death-rate, is sometimes represented by the letter r and determines the specific rate of growth (or, if the death-rate should exceed the birth-rate, decline) of the population (see p. 86); using the r designation, equation 7.2 becomes

$$N_{t+1} = N_t + rN_t\,\Delta t \tag{7.3}$$

which may be rearranged (by subtracting N_t from both sides and dividing through by Δt) to produce

$$\frac{N_{t+1} - N_t}{\Delta t} = rN_t.$$

The term $(N_{t+1} - N_t)$ is the change in population between time $t+1$ and time t; this can be more concisely expressed by ΔN, where the Greek capital delta, Δ, as in the term Δt, indicates a change in the variable. Equation 7.3 is now

$$\frac{\Delta N}{\Delta t} = rN_t. \tag{7.4}$$

Notice that in the left-hand term, the change in the population is calculated over the time interval Δt, so it would be measured as, say, persons per year; in other words, it is the rate of change of the population during the time interval. A year is a time interval of finite length. It is sometimes useful to consider the change in population over an infinitesimally small time interval—that is, at an instant—and thus derive the instantaneous rate of population change. When instantaneous rates are studied, the letter d is used to indicate a change rather that Δ; the d can then be thought of as denoting a tiny change. Using the d notation, equation 7.4 is

$$\frac{dN}{dt} = rN. \tag{7.5}$$

As we saw earlier (p. 85), the solution of this equation is

$$N_t = N_0 e^{rt} \tag{7.6}$$

where N_0 is the initial population. Depending on the sign of r, equation 7.6 describes exponential growth or decline; in essence, this is Quetelet's model.

Equation 7.5 assumes unlimited growth. More realistically, an inhibiting factor can be added which reduces the instantaneous rate of growth in proportion to the square of the population size:

$$\frac{dN}{dt} = rN - cN^2 \tag{7.7}$$

where the parameter c determines the reduction in growth rate per unit of population. Defining an upper limit to which the population may grow (the carrying capacity of the environment, K) in such a way that $c = r/K$, we have

$$\frac{dN}{dt} = rN - \frac{r}{K}N^2. \tag{7.8}$$

The solution of equation 7.8 is

$$N = K/(1 + Ce^{-rt}) \tag{7.9}$$

which, with C a constant related to the initial population, describes a logistic growth curve, of the kind first developed by Verhulst and Pearl, for positive values of C (see p. 86).

These basic ideas can be extended to models of inter-regional population growth, models in which population growth is studied in a set of regions, migration between each of which may take place. A simple model for two regions—California and The Rest of the United States is described by Rogers (1968). We shall use Rogers's example though the model formulation will differ: Rogers's method of analysis, which used matrix algebra, will be briefly discussed. The state variables and parameters used in the model are listed in Table 7.3.

The initial state of the system is given by the populations of the two regions in 1955. At that date, the population of California, $N_{1,t}$, the subscript 1 signifying California, was 12 988; and the population of The Rest of the United States, $N_{2,t}$, the subscript 2 signifying The Rest of the

Table 7.3
Symbols and values of state variables and parameters in the California–Rest of the United States population model

State variables and parameters	California		The Rest of the United States	
	Symbol	Value	Symbol	Value
Population at time t, 1955 (1000s)	$N_{1,t}$	12 988	$N_{2,t}$	152 082
Population at time $t+1$, 1960 (1000s)	$N_{1,t+1}$	?	$N_{2,t+1}$	?
Specific growth rate (per 1000 people per 5 years)	r_1	0.0215	r_2	0.0667
Specific migration rate (per 1000 people per 5 years)	m_1	0.0127	m_2	0.0627

United States, was 152 082, both figures being given in thousands of people. By the same arguments used in building equation 7.3, and here including a migration component-of-change, the state of the system after a five-year period (the model was calibrated by Rogers for this time interval) is defined, with the Δt term that appears in equation 7.3 omitted because its value is one, as

$$N_{1,t+1} = N_{1,t} + r_1 N_{1,t} + m_2 N_{2,t}$$
$$N_{2,t+1} = N_{2,t} + r_2 N_{2,t} + m_1 N_{1,t}$$

or, what is the same thing

$$\begin{aligned} N_{1,t+1} &= (1+r_1)N_{1,t} + m_2 N_{2,t} \\ N_{2,t+1} &= (1+r_2)N_{2,t} + m_1 N_{1,t}. \end{aligned} \tag{7.10}$$

In these equations, $m_1 N_{1,t}$ is the migration from California to The Rest of the United States, and $m_2 N_{2,t}$ is the migration from The Rest of the United States to California. Substituting the values of the terms which appear on the right-hand side of equations 7.10 from Table 7.3 we end up with

$$N_{1,t+1} = (1.0215)(12\,988) + (0.0127)(152\,082) = 15\,199$$
$$N_{2,t+1} = (1.0667)(152\,082) + (0.0627)(12\,988) = 163\,040.$$

The model thus predicts that by 1960 the state of the system has changed: the population of both regions has grown, in California to 15 199 thousands and in The Rest of the United States to 163 040 thousands. The predicted 1960 populations, $N_{1,t+1}$ and $N_{2,t+1}$, could then be fed into the following equations to find the population in each region in 1965, that is, $N_{1,t+2}$ and $N_{2,t+2}$

$$N_{1,t+2} = (1.0215)N_{1,t+1} + (0.0127)N_{2,t+1}$$
$$N_{2,t+2} = (1.0667)N_{2,t+1} + (0.0627)N_{1,t+1}.$$

This iterative procedure can be repeated indefinitely into the future.

It is interesting to arrange the parameters of the model as a table:

$$\begin{bmatrix} r_1 & m_1 \\ m_2 & r_2 \end{bmatrix}$$

and with values put in

$$\begin{bmatrix} 1.0215 & 0.0127 \\ 0.0627 & 1.0667 \end{bmatrix}.$$

This table, or matrix, determines the growth of the system. It is not surprising, therefore, that by manipulating the growth-parameter matrix mathematically, several important growth characteristics of the system may be ascertained. For instance, a number called the dominant latent root of the matrix can be extracted—it is 1.0802. The significance of the latent root is that, if, as in our example, it exceeds 1.0, the system is capable of growing; if it is equal to 1.0, the system state remains the same; if it is less than 1.0, the system will decline.

The over-all specific rate of growth of the system, r, at stability is defined as the natural logarithm of the dominant latent root

$$r = \ln(1.0802)$$
$$= 0.07696.$$

The time taken for the population to double itself, T, is defined by

$$T = (\ln 2)/r$$
$$= 0.69315/0.07696 \text{ per 5 years}$$
$$= 45 \text{ years approximately.}$$

Associated with the dominant latent root of the growth matrix is a system state which is stable. The system is growing so the stable state means that the relative proportions of the populations of California and The Rest of the United States to the total United States population remains constant. By a series of calculations, these proportions are found to be 0.1778:0.8221, California being the smaller value.

7.2.2. *Cohort-survival models*

Leslie's (1945) original cohort-survival model predicts the age structure of a female population after a given interval of time, given the age structure at the start of the time interval and specific birth- and survival-rates for each age group. The model disaggregates a population not by regions, but by age groups and could be formulated as a set of equations similar to equations 7.10 but without a migration component and with $N_{1,t}$ being the population of the first age group at time t and $N_{2,t}$ being the population of the second age group at time t, and so on.

Disaggregated in this way, components-of-change models can be useful tools in studying populations, far more so than the somewhat crude exponential and logistic models. Usher (1972), for example, used a version of the Leslie model to investigate the effect of over-exploitation on the blue whale (*Balaenoptera musculus*) population. The model is based on data for the blue whale collected in the 1930s, before the virtual extinction of the blue-whale population in the early 1940s. To seek the reasons for the near-catastrophic decline in the blue-whale population, age-specific population parameters—birth-rates and survival-rates—were established for each of seven age groups (0 to 1, 2–3, 4–5, 6–7, 8–9, 10–11, and 12 years plus). The parameter values for each age group are listed in Table 7.4.

The birth terms (or technically speaking in this study the fecundity terms) indicate that female blue whales do not breed in the first four years of life and that at full breeding each cow produces one calf every two years (since the sex ratio of the population is 1:1 this calf has a 0.5 chance of being female). The survival terms, 0.87, are based on the best available estimate of natural mortality. Because all the whales in the last age group do not die in a two-year period, some indeed live to be 40, the survival value of 0.8 has been placed in the bottom right-hand corner of the matrix; this gives whales in the oldest age-class a life expectancy of 7.9 years. The parameters shown in Table 7.4, like the parameters in the two-region population model (p. 152), can be arranged as a growth matrix from which the growth characteristics of the blue-whale population may be extracted. The dominant latent root of the blue-whale population's growth matrix is 1.0986, which shows that the population is capable of growing. The intrinsic rate of natural increase of the blue-whale population is given by the natural logarithm of the dominant latent

Table 7.4
Parameters for a cohort-survival model of the blue whale population

Parameters (per cow per two years)	*Age groups*						
	0–1	2–3	4–5	6–7	8–9	10–11	12^{+}
Birth-rate	0	0	0.19	0.44	0.50	0.50	0.45
Survival-rate	0.87	0.87	0.87	0.87	0.87	0.87	0.80

root (p. 152) and is 0.094. We may denote the dominant latent root as λ_1. Now λ_1 can be used to estimate the harvest which can be taken without causing a decline in the population. The population size increases from N to $\lambda_1 N$ over a two-year period so the harvest which can be taken, H, expressed as a percentage of the total population, is

$$H = 100\left(\frac{\lambda_1 - 1}{\lambda_1}\right).$$

This is roughly 4.5 per cent of the total population per year. If this harvest rate were exceeded, the population would decrease unless homeostatic mechanisms came into play and altered the fecundity and survival-rates. Evidence suggests that when under pressure, the blue-whale population shows a slight increase in the pregnancy rate and individuals reach sexual maturity earlier in life and may generally grow more rapidly. The question is, can these responses counterbalance the effects of exploitation? For a definite answer to this question long-term studies of data collection and simulation with mathematical models are needed.

The cohort-survival and multi-regional population models have been combined by Rogers (1968). The combined model, if formulated in the same vein as equations 7.10, would have one equation for each age group in each region. With just ten age groups and ten regions this would mean 100 equations which would be cumbersome to handle. For this reason multi-regional cohort-survival models are best tackled in matrix form but, because the notation used is likely to befuddle the uninitiated reader, their formulation will not be pursued here. The original model given by Rogers (1968) can be refined: population may be subdivided into male and female components and migration models can be used to estimate the rates required in the migration matrices; these refinements, their assumptions, and the means by which they are fitted into the model are outlined by Wilson (1974, pp. 84–7). Additionally, the model has been developed in three ways. Firstly, Wilson (1974) has shown how allowance can be made for varying age-group intervals and varying time intervals (projection periods). Secondly, Wilson (1974) shows how time can be considered as a continuous variable. And thirdly, Rees and Wilson (1977) show how, so as to place population models on a firm analytical footing, an accounting framework may be devised in which to develop the model and in which to facilitate certain rate estimations from commonly available population data.

7.2.3. *Interacting populations*

In nature, the growth of one population is seldom independent of the growth of other populations living in the same environment: populations interact. Two types of interaction have been modelled—competition and predation. The theoretical basis of competition was established by Lotka (1925) and Volterra (1926) and has reached a fairly sophisticated level (see Maynard Smith, 1974, pp. 59–68). However, competition models do not have a locational component and need not concern us here. Of more interest are predator–prey systems which, although originally modelled by Volterra (1926) on a non-spatial basis, have recently been extended to consider the inter-regional movements of predators and prey.

The basic ideas of predator–prey interactions are straightforward and run as follows. Consider the hypothetical case of a population of weasels, N_2, the predators, and a population of rabbits, N_1, the prey. If the rabbit population is large then, with an abundance of food around, the weasel population will flourish and increase. But as the weasel population increases, so will the rabbit population diminish. They prey shortage will cause a decline in the weasel population and, enjoying the dearth of predators, the rabbit population will increase. And so the cyclical change in rabbit and weasel populations will continue. The equations which describe these changes can be expressed in several ways; Volterra's equations were

$$N_{1,t+1} = N_{1,t} + (r_1 N_1 - p_1 N_1 N_2)\,\Delta t$$

$$N_{2,t+1} = N_{2,t} + (-d_2 N_2 + p_2 N_1 N_2)\,\Delta t$$

where $r_1 N_1$ describes the exponential growth of the prey population; $N_1 N_2$ represents the contacts between prey and predators; p_1 and p_2 are predator constants, p_1 being the prey death-rate due to predators, and p_2 being the per capita rate of increase of the prey population; and $d_2 N_2$ represents the specific death-rate of the predator population. The solutions to these equations are, as we have seen (p. 4), cyclical, but the cycles are not necessarily stable. Stability obtains where the predator is inefficient at catching its prey (Figure 1.2, p. 5).

Maynard Smith (1974, pp. 72–83) developed a model which allows the prey and predator populations to migrate from one region to another. An area of interest is divided into square cells. Each cell may exist in one of eight states: empty; few prey; increasing prey; many prey; many prey, few predators; many prey, increasing predators; many prey, many predators; and, few prey, many predators. The possible transitions from one state to another are shown in Figure 7.10. The state transitions are expressed as probabilities for prey and predators, several migration assumptions being possible. A number of computer simulations of the model led Maynard Smith (1974, p. 79) to conclude that coexistence of prey and predator populations is rather easily attained and is favoured by prey with a high capacity for migration, cover or refuge for the prey, the migration of predators during a restricted period, and a large number of cells. If the

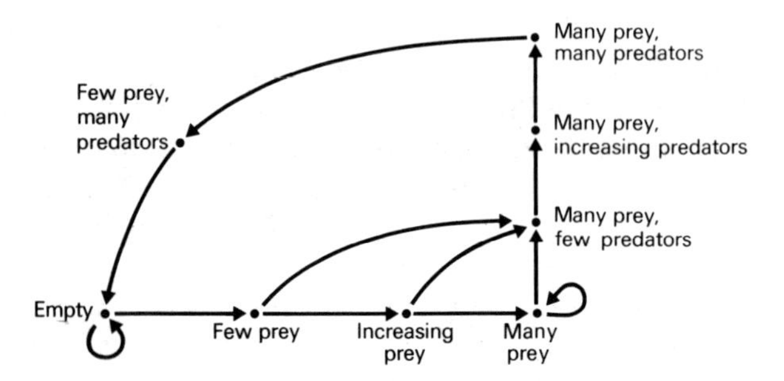

FIG. 7.10. Possible state transitions in Maynard Smith's model. Reprinted with permission, from J. Maynard Smith (1974), *Models in ecology*, Cambridge University Press, London.

predators' ability to migrate is too low, they will become extinct, if it is high, coexistence is possible so long as the prey are equally mobile. Interestingly, under the assumption that predators migrate only when they are hungry, long-term oscillations occur in the number of cells containing predators, probably because of a delay effect arising from the time elapsing between the moment a cell is invaded by prey or predators and the time when that cell becomes a source of invaders.

7.3. Spatial interaction models

Given a set of zones or sub-regions, a family of models, deterministic in character, may be employed to analyse the interaction between zones as reflected in migration patterns, shopping behaviour, journey-to-work trips, and the like. Several possible zoning systems for the same region may be adopted, including wards within a local authority, subdivision of wards such as Census Enumeration Districts, postal districts, or regular grid squares (Figure 7.11). Interest will normally focus on one state variable, say population. The population in each zone may be denoted by the symbol P_i where $i = 1, n$ identifies a particular zone in a region of n zones. Next a variable is needed to define interaction between zones. Say we are interested in the number of trips between zones then we may define a variable T and give it two subscripts, i and j, one for trip origin and one for trip destination. T_{ij} is thus the number of trips between zone i and zone j, and specifically T_{36} would be the number of trips between zone 3 and zone 6. If there are n zones then there will be $n \times n$ trip variables and these could be represented as a trip matrix in which row headings are zones of origin and column headings are zones of destination (Table 7.5). The row totals, which we shall call O_i $(i = 1, 3)$, are the total number of trips originating in zone i and the column totals, which we shall call D_j $(i = 1, 3)$, are the total number of trips ending in zone j. The grand total of either row *or* column sums is the total number of trips in the

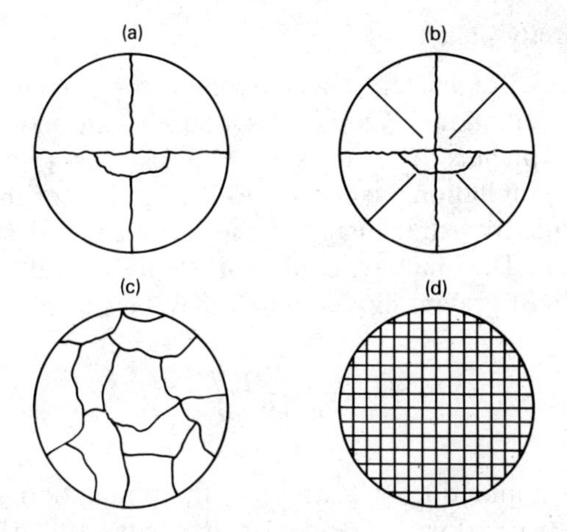

FIG. 7.11. Spatial zoning systems. Reprinted with permission, from A. G. Wilson and M. J. Kirkby (1975), *Mathematics for geographers and planners*, Oxford University Press, Oxford.

region, worked out as total number of trip origins, T, or trip destination, T'. In other types of interaction it will be necessary to define other variables which relate to pairs of zones: d_{ij} is the distance between zone i and zone j; t_{ij} is the travel time between them; and c_{ij} is the cost of travel between them. With this basis, we can now consider the main members of the family of spatial interaction models, by name, the grandfather of them all the simple gravity model, the shopping model, and the journey-to-work model.

Table 7.5
A trip matrix for three zones

		to zone j (destination) $j = 1, 3$ 1	2	3	*row totals*
from zone i (origin) $i = 1,3$	1	T_{11}	T_{12}	T_{13}	$\sum_{j=1}^{3} T_{1j} = O_1$
	2	T_{21}	T_{22}	T_{23}	$\sum_{j=1}^{3} T_{2j} = O_2$
	3	T_{31}	T_{32}	T_{33}	$\sum_{j=1}^{3} T_{3j} = O_3$
Column totals		$\sum_{i=1}^{3} T_{i1} = D_1$	$\sum_{i=1}^{3} T_{i2} = D_2$	$\sum_{i=1}^{3} T_{i3} = D_3$	$\sum_{i=1}^{3} O_i = T; \sum_{i=1}^{3} D_j = T'$

7.3.1. *The gravity model*

As early as 1885, Ravenstein had embodied the principles of Newton's Law of mass interaction, which states that the attraction between two masses—atoms, planets or stars, and the like—is proportional, by a constant *G*, the gravitational constant, to the product of the masses of the two masses divided by the square of the distance separating them, into laws of migration. The analogy, implicit in Reilly's (1929) work, was used in Stewart's (1948) gravity model which may be put as follows

$$F_{ij} = a\frac{P_i P_j}{d_{ij}^b} \qquad (7.11)$$

where P_i is the population of town i; P_j the population of town j; d_{ij} is the distance between towns i and j; a is a constant of proportionality analogous to the gravitational constant but used as a fudge factor to ensure that predicted interaction between towns i and j, F_{ij}, roughly matches the observed movements; and b is a distance–decay exponent which varies with different types of movement, between areas, and over time and, with the a value, may be estimated empirically by studying situations in which all other terms in equation 7.11 are known.

By way of example, we shall consider Keeble's (1971) study of the movement of industry, as measured by number of jobs created, from the south-east of the United Kingdom to its peripheral regions between 1945 and 1965. Movement of industry in this study is unidirectional and the gravity model was used in the form

$$F_{ij} = a\frac{M_j}{d_{ij}^b}$$

where F_{ij} is the number of jobs created; M_j is the number of unemployed in peripheral region j in the middle of the period (1954), that is, a measure of the attractiveness of the region to mobile industry; d_{ij} is the straight line distance between the centres of the regions. It was found that the best-fit value of the distance exponent, b, was 2, and the value of a, the term which scales predicted movements to observed ones, was 53 000. The results are shown in Table 7.6.

The simple gravity model is very crude and has been subjected to several refinements (see Isard *et al.*, 1960). The gravitational concept has also permeated other fields of geography. For instance, in estimating the potential flows of say population from one region to another, a population potential surface can be constructed (see p. 54) and this engages gravity concepts.

Table 7.6

Movement of industry from the South-East region to peripheral regions

Region	M_j *1954 unemployed workforce*	d_{ij} *distance between regions (miles)*	F_{ij} *predicted movements of jobs*	*observed movements of jobs*
Northern Ireland	33 000	320	17 079	16 900
Scotland	59 500	360	24 332	24 500
Northern	28 300	250	23 998	35 800
Wales	22 900	140	61 923	43 200
Merseyside	18 900	180	30 916	36 000
Devon and Cornwall	8 600	190	12 625	10 600

Based on data in D. E. Keeble (1971).

7.3.2. *The shopping model*

This model was formulated by Huff (1964) and considered by Lakshmanan and Hansen (1965); a detailed account of it is to be found in Wilson (1974). The purpose of the model is, essentially, to predict the spatial distribution of shopping sales and is thus, unlike the journey-to-work and basic gravity model, a location model and not solely an interaction one. In outline the model considers the cash flow, S_{ij}, between residences in zone i and shops in zone j in relation to: the population of zone i, P_i; the mean expenditure on shopping goods per person in zone i, e_i; the attractiveness of shops in zone j, W_j; and the cost of travel from zone i to zone j, c_{ij}. The following relations are usually assumed: the flow of cash is proportional to the total spending power in zone i, the attractiveness of shops, and a decreasing function of travel cost between zones. We may put these ideas as symbols, with $\propto$ meaning proportional to, as

$$\begin{aligned} S_{ij} &\propto e_i P_i \\ S_{ij} &\propto W_j^{\alpha} \\ S_{ij} &\propto c_{ij}^{-\beta} \ (\beta > 0) \end{aligned} \tag{7.12}$$

which, with K as a constant of proportionality, implies that

$$S_{ij} = K(e_i P_i) W_j^{\alpha} c_{ij}^{-\beta}. \tag{7.13}$$

Equation 7.13 represents a simple model, akin to Reilly's (1929) gravity model.

The total cash flow out of zone i, which is given by $\sum_{j=1}^{n} S_{ij}$, must be equal to the total expenditure on shopping goods in zone i, which is given by e_iP_i; therefore

$$\sum_{j=1}^{n} S_{ij} = e_iP_i. \tag{7.14}$$

In practice, the constant of proportionality, K, may be calculated for each zone so that equation 7.14 is satisfied: we thus redefine K as K_i. So

$$S_{ij} = K_i(e_iP_i)W_j^{\alpha}c_{ij}^{-\beta}. \tag{7.15}$$

Substituting into equation 7.14 and solving for K_i yields

$$K_i = 1\Big/\sum_{j=1}^{n} W_j^{\alpha}c_{ij}^{-\beta} \tag{7.16}$$

Equations 7.15 and 7.16 are the full statement of the model though 7.16 could be substituted for K_i in 7.15 to give

$$S_{ij} = (e_iP_i)\frac{W_j^{\alpha}c_{ij}^{-\beta}}{\sum_{j=1}^{n} W_j^{\alpha}c_{ij}^{-\beta}}. \tag{7.17}$$

This model predicts that cash will flow between all pairs of zones in the region, though the flow could be very small if e_iP_i is very small, W_j is very small, or c_{ij} is very large (and hence $c_{ij}^{-\beta}$ is very small). Thus only a small number of zones in the study area may in practice contribute to total shopping sales in any one zone. Thus the model could incorporate the definition of market areas.

Wilson and Kirkby (1975, pp. 95–7) give a hypothetical example of the model for a three-zone spatial system. The cost of travelling between and within zones and other relevant data are shown in Table 7.7.

Table 7.7
Data for shopping model in a three-zone system

Zone	c_{ij}						
	1	2	3	e_i	P_i	e_iP_i	W_j
1	1.0	5.0	5.0	2	50	100	10
2	5.0	2.585	5.0	1	1000	1000	100
3	5.0	5.0	2.0	1	500	500	20

α and β are both assumed to be 1.0. In the travel-cost matrix (the c_{ij}s), the main diagonal elements are within-zone travel costs and the off-diagonal elements are the between-zone travel costs. From data and tables in Wilson and Kirkby (1975).

Table 7.8
Predicted cash flows between zones

	to zone		
	1	2	3
from 1	29.4	58.8	11.8
zone 2	44.8	864.0	89.6
3	31.3	313.0	156.5
column totals	105.5	1235.8	257.9

From data in Wilson and Kirkby (1975).

From this information, and using equation 7.16, the values of the K_is may be calculated: they are $K_1 = 1/34$, $K_2 = 1/44.6$ and $K_3 = 1/32$. Substituting these values and other relevant data from Table 7.7 into equation 7.15 enables the set of cash flows between each pair of zones to be evaluated; for instance, S_{12}, the flow between zone 1 and zone 2, is

$$S_{11} = K_1(e_1P_1)W_1c_{11}^{-1} = (1/34)\times 1000 = 29.4.$$

The complete model output is shown in Table 7.8; notice the column totals give the total predicted sales in each shopping centre.

The chief feature of the shopping model is that the only constraint on flows is the flow totals as defined by equation 7.14; it is therefore known as a singly constrained or production-constrained model and may be contrasted with doubly constrained journey-to-work model which we shall now consider.

7.3.3. *The journey-to-work model*

The model is set up with variables as defined in Table 7.5 (p. 157) in which T_{ij} is the number of work trips from zone i to zone j in a spatial system; O_i is the total number of work trip origins in zone i; and D_j is the total number of work-trip destinations (and number of jobs) in zone j. The constraints on the model are that

$$\sum_j T_{ij} = O_i \tag{7.18}$$

that is, the total predicted number of trips starting from zone i must be equal to the number of workers resident in that zone;

$$\sum_i T_{ij} = D_j \tag{7.19}$$

that is, the total number of trips ending in zone j must be equal zone to the number of jobs available in that zone. By the same arguments (equations 7.12) as in the shopping model we may assume that

$$T_{ij} \propto O_i \qquad T_{ij} \propto D_j \quad \text{and} \quad T_{ij} \propto c_{ij}^{-\beta} \qquad (7.20)$$

where c_{ij} is the cost of travel between zones i and j. To ensure the two constraints are satisfied, a double set of proportionality constants, equivalent to K_is in the shopping model, are required; these are denoted as A_iB_j. So we have

$$T_{ij} = A_iB_jO_iD_jc_{ij}^{-\beta}. \qquad (7.21)$$

Substituting equation 7.21 into equation 7.18 and rearranging gives

$$A_i = 1/\sum_j B_jD_jc_{ij}^{-\beta}. \qquad (7.22)$$

Similar, substituting equation 7.21 into equation 7.19 and rearranging gives

$$B_j = 1/\sum_i A_iO_ic_{ij}^{-\beta}. \qquad (7.23)$$

The full model is given by equations 7.21, 7,22 and 7.23. To solve for A_i and B_i an iterative procedure is used because equations 7.22 and 7.23 are non-linear (they contain a reciprocal), simultaneous equations.

7.3.4. *The full family of interaction models*

In general, the basic interaction model equation takes the form

$$\text{Interaction} = \text{constant} \times \text{mass} \times \text{mass} \times \text{distance function}$$

and is subject to four main types of constraint as shown in Table 7.9. In the unconstrained case, measures of attractiveness in both origin and destination zones are employed—both sets of interaction sums, O_i and D_j,

Table 7.9
Main models of spatial interaction

Case	*Interaction*	*Constants*	*Mass*	*Mass*	*Distance Function*
Unconstrained	T_{ij}	K	$W_i^{(1)}$	$W_j^{(2)}$	$f(c_{ij})$
Production-constrained	T_{ij}	$K(=K_i)$	O_i	$W_j^{(2)}$	$f(c_{ij})$
Attraction-constrained	T_{ij}	B_j	$W_i^{(1)}$	D_j	$f(c_{ij})$
Production-attraction-constrained	T_{ij}	A_iB_j	O_i	D_j	$f(c_{ij})$

are unspecified as in the simple gravity models. In the production-constrained case the total production of flows out of zone i, O_i, is given; the shopping model is an example of this. In the attraction-constrained case the total number of flows into zone j, D_j, is given; this case is the mirror image of the production-constrained case. In the production-attraction-constrained case both the O_is and D_js are specified as in the journey-to-work model.

7.4. Stochastic models

A variety of stochastic models are used to tackle geographical problems: we shall consider two types which lie within the ambits of systems analysis and can be used in a regional context, namely, Markov-chain models and Monte Carlo models.

7.4.1. *Markov-chain models*

To show how Markov-chain models are used we shall develop the example of daily weather changes mentioned in §5.3 in which a transition-probability matrix, **P**, was set up to define the probability of a fine day being followed by a fair one, and so on, as follows.

$$\begin{array}{cc} & \begin{array}{ccc} \textit{fine} & \textit{fair} & \textit{foul} \end{array} \\ \mathbf{P} = \begin{array}{c} \text{fine} \\ \text{fair} \\ \text{foul} \end{array} & \begin{pmatrix} 0.3 & 0.2 & 0.5 \\ 0.4 & 0.4 & 0.2 \\ 0.3 & 0.6 & 0.1 \end{pmatrix} \end{array}.$$

In the general case for a 3×3 matrix the elements will be written

$$\mathbf{P} = \begin{pmatrix} p_{11} & p_{12} & p_{13} \\ p_{21} & p_{22} & p_{23} \\ p_{31} & p_{32} & p_{33} \end{pmatrix}.$$

The information in the **P** matrix can be used to discover the probability of changing between any two states over one or two days given a starting state. Take, for example, a fine day as a starting-point. A tree diagram may be constructed which shows transitions to other states after one and two days (Figure 7.12); this shows that there are three different ways of going from fine to foul in two days. The probability of changing from fine to foul in two days, which may be symbolized as p_{13}^2, is

$$p_{13}^2 = (p_{11}p_{33} + p_{12}p_{23} + p_{13}p_{33}).$$

Substituting actual values for the ps we have

$$\begin{aligned} p_{12}^2 &= (0.3 \times 0.5) + (0.2 \times 0.2) + (0.5 \times 0.1) \\ &= 0.24. \end{aligned}$$

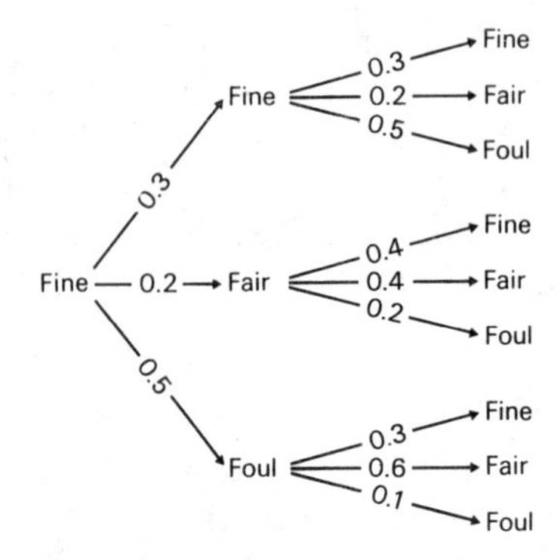

FIG. 7.12. A tree diagram showing transitions to other states over two days starting with a fine day.

The probabilities of changing between all other pairs of states over two days can be evaluated in the same way. The general formula for n days (time steps) is

$$p_{ij}^{n} = \sum_{k=1}^{n} p_{ik}p_{kj}$$

where k is the number of different paths which may be taken to get from one state to another in two days.

A more efficient way of doing the same thing is to raise the transition matrix to successively higher powers. For instance, to find p_{ij}^{2} we simply calculate $\mathbf{P}^2$, that is multiply $\mathbf{P}$ by itself[1]

$$\mathbf{P}^2 = \begin{bmatrix} 0.3 & 0.2 & 0.5 \\ 0.4 & 0.4 & 0.2 \\ 0.3 & 0.6 & 0.1 \end{bmatrix} \begin{bmatrix} 0.3 & 0.2 & 0.5 \\ 0.4 & 0.4 & 0.2 \\ 0.3 & 0.6 & 0.1 \end{bmatrix}$$

$$= \begin{bmatrix} 0.32 & 0.44 & 0.24 \\ 0.34 & 0.36 & 0.30 \\ 0.36 & 0.36 & 0.28 \end{bmatrix}.$$

[1] The way to do this is to evaluate

$$\mathbf{P}^2 = \begin{bmatrix} (p_{11}\times p_{11})+(p_{12}\times p_{21})+(p_{13}\times p_{31}) & (p_{12}\times p_{12})+(p_{12}\times p_{22})+(p_{13}\times p_{32}) & (p_{11}\times p_{13})+(p_{12}\times p_{23})+(p_{13}\times p_{33}) \\ (p_{21}\times p_{11})+(p_{22}\times p_{21})+(p_{23}\times p_{31}) & (p_{21}\times p_{12})+(p_{22}\times p_{22})+(p \times p_{32}) & (p_{21}\times p_{13})+(p_{22}\times p_{23})+(p_{23}\times p_{33}) \\ (p_{31}\times p_{11})+(p_{32}\times p_{21})+(p_{33}\times p_{31}) & (p_{31}\times p_{12})+(p_{32}\times p_{22})+(p_{33}\times p_{32}) & (p_{31}\times p_{13})+(p_{32}\times p_{23})+(p_{33}\times p_{33}) \end{bmatrix}$$

For instance, the first element of $\mathbf{P}^2$ is $(0.3\times 0.3)+(0.2\times 0.4)+(0.5\times 0.3)=0.32$.

The elements of matrix $\mathbf{P}^2$ are the probabilities of passing between pairs of states over two days. Similarly $\mathbf{P}^3$, which is of course evaluated as $\mathbf{P}^2 \times \mathbf{P}$, gives the probabilities of passing between pairs of states after three days; we have

$$\mathbf{P}^3 = \begin{bmatrix} 0.3384 & 0.3856 & 0.2760 \\ 0.3392 & 0.3872 & 0.2736 \\ 0.3384 & 0.3888 & 0.2728 \end{bmatrix}.$$

The process of powering a probability matrix leads eventually to a matrix which does not change on being raised to a higher power. In our example this occurs with matrix $\mathbf{P}^4$ which, with the numbers rounded, is

$$\mathbf{P}^4 = \begin{bmatrix} 0.3387 & 0.3871 & 0.2742 \\ 0.3387 & 0.3871 & 0.2742 \\ 0.3387 & 0.3871 & 0.2742 \end{bmatrix}.$$

Notice that the rows in the $\mathbf{P}^4$ matrix are the same and give the equilibrium proportions of the different states. This is an important property of the matrix and means that the probability of passing to any state is independent of the starting state. Given a sufficient number of days, the row elements in the $\mathbf{P}^4$ matrix are the probabilities of changing states; this implies that the occurrence of states (weather conditions) will also be in these proportions over a large number of days.

A Markov-chain model has been used by Waggoner and Stephens (1970) to study succession in a North American mixed hardwood forest. For 327 plots, the dominant tree species—maple, oak, birch, other major hardwood species, and minor species—were recorded and transition probabilities between them calculated for the decade 1927 to 1937 (Table 7.10).

Table 7.10
Transition probabilities for mixed hardwood forest

			1927		
1937	*Maple*	*Oak*	*Birch*	*Other*	*Minor*
Maple	0.82	0.16	0.13	0.07	0.07
Oak	0.07	0.72	0.02	0.03	0.07
Birch	0.02	0.08	0.83	0.07	0.07
Other	0.0	0.0	0.02	0.69	0.07
Minor	0.09	0.04	0.0	0.14	0.72

Based on Waggoner and Stephens (1970).

In matrix form, Table 7.10 can be used to predict changes of state in the 327 forest plots. Waggoner and Stephens (1970) found good agreement between predicted and observed states of the forest to 1967.

7.4.2. *Monte Carlo models*

Monte Carlo methods have been applied to a variety of geographical diffusion problems. For instance, Morrill (1965) was able to simulate the expansion of the Negro ghetto in Seattle between 1940 and 1960. At a larger scale, Levison *et al.* (1973) simulated the ancient colonization of the Polynesian Islands. The aim of this study was to help resolve the archaeological debate concerning the relative importance of random-drift voyaging and the navigational skills of settlers on the sequence of island occupation. As an example of a Monte Carlo model we shall consider Hägerstrand's (1967) model of the spread of farm subsidies in the Asby area of southern Sweden.

A farm subsidy was introduced by the Swedish government in 1928 to encourage farmers with small farms (less than 8 hectares of tilled land) to improve and enclose pastures and end the practice of allowing cattle to forage in open woodland during summer months. To simulate the spread of the farm subsidy after 1928, Hägerstrand divided the Asby region into a lattice of 125 square cells each covering 5×5 km (Figure 7.13). The

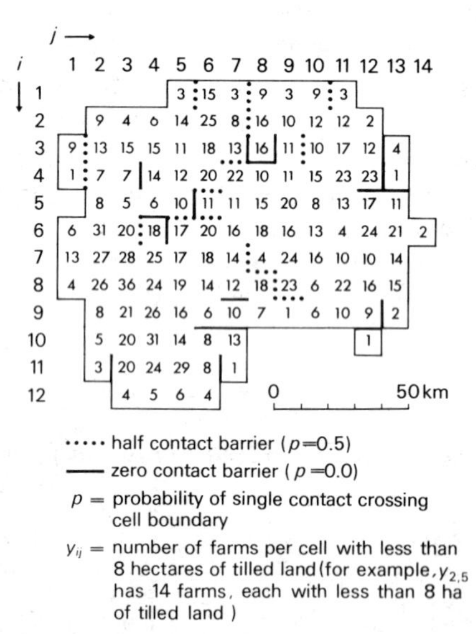

FIG. 7.13. Model environment representing part of southern Sweden in 1929. From T. Hägerstrand (1967), 'On Monte Carlo simulation of diffusion', in *Quantitative geography*, Part I: *Economic and cultural topics* (ed. W. L. Garrison and D. F. Marble), pp. 1–32 (Northwestern Studies in Geography No. 13), Northwestern University Press, Evanston, Illinois.

following variables were then defined for each cell: the total number of farms in each cell entitled to receive the subsidy, y_{ij} (Figure 7.13); and the number of farmers in each cell who have adopted the subsidy by a particular time, x_{ijt}; The model, from the initial distribution of adopters, x_{ijo} (Figure 7.14a), predicts future patterns of adopters according to

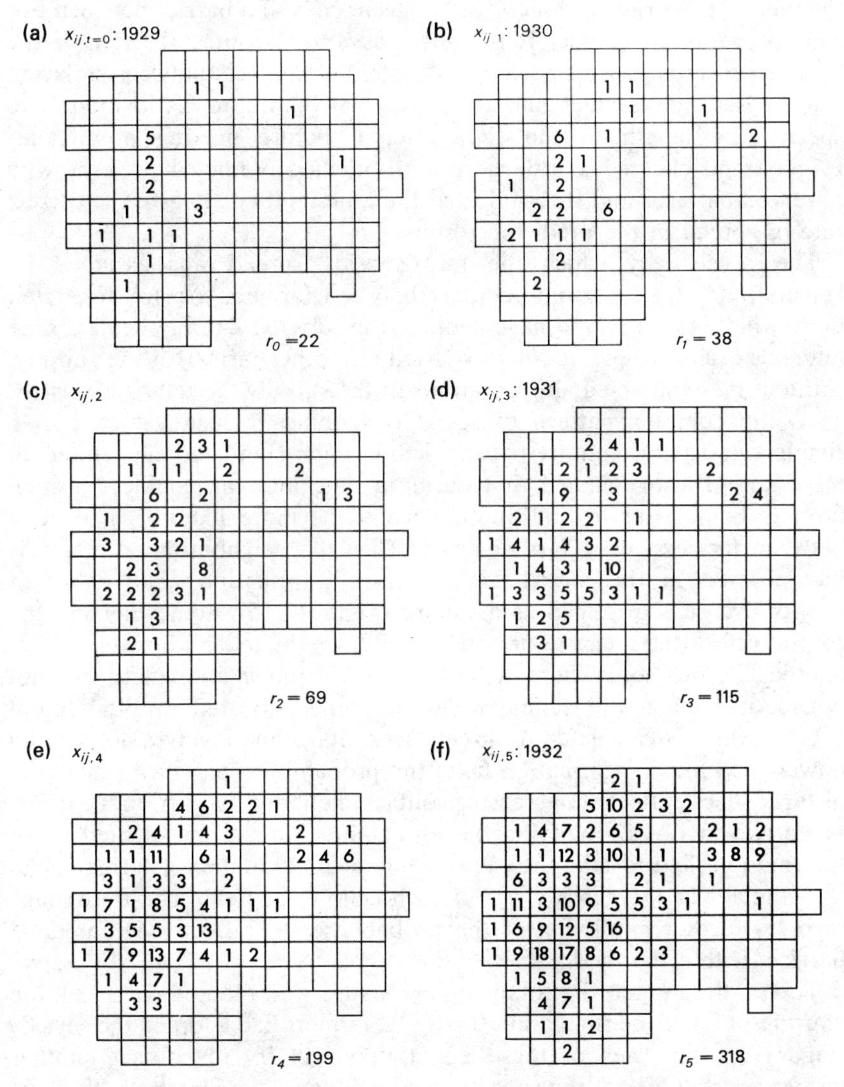

FIG. 7.14. One simulation of farm-subsidy diffusion. From T. Hägerstrand (1967), 'On Monte Carlo simulation of diffusion', in *Quantitative geography*, Part I: *Economic and cultural topics* (ed. W. L. Garrison and D. F. Marble), Northwestern University Press, Evanston, Illinois.

certain assumptions. The assumptions concern the environment, time, and the processes which bring about the spread of subsidy adoption during a time interval.

Environmental factors, such as the presence of fen, forest, and lake, which will obstruct the spread of subsidy adoption can be represented in the model as barriers between two adjacent cells. If a barrier between two cells prevents all contact it is known as a zero-contact barrier; if the barrier prevents just 50 per cent of potential contacts between adjacent cells it is a half-contact barrier. It is usual in Monte Carlo models to specify the length of the time interval before making predictions. Hägerstrand cheated a little here and matched predicted patterns with observed patterns and then adjusted the time intervals to give the correct rate of spread in the predicted results.

The process by which the farm-subsidy spread was assumed by Hägerstrand to be personal contact between farmers. At any time step there will be farmers who have adopted the subsidy. During the next time interval, each of these adopters will tell one other farmer of the subsidy. If the farmer contacted already knows of the subsidy the telling is wasted. Thus the complex pattern of social communications is reducted to a simple one-to-one contact pattern within each simulated time interval. Hägerstrand also assumed that during a time interval, contact between farmers living near to one another would be more likely than contact between farmers living farther apart. This, the neighbourhood effect, is incorporated in the model as the mean information field, a set of twenty-five cells arranged as a square (Figure 7.15). Somewhere in the central cell of the mean information field lines the teller who will contact another farmer somewhere within the mean information field but not outside it. Each cell in the mean information is allocated a probability of contact which is calculated in three stages. Stage one involves defining an unweighted mean information field, the probabilities in which reflect the distance–decay effect of decreasing contact likelihood away from the teller. Stage two involves modifying the probabilities in the unweighted mean information field to take into account the actual number of farmers who are eligible for the subsidy, y_{ij}, in each cell of the field. The third stage involves further modifying of the probabilities to reflect environmental barriers within the field.

To run the model, the following procedure was adopted. In 1929, the starting-point of the simulation ($t = 0$), 22 farmers had adopted the subsidy (Figure 7.14a). Each of these 22 farmers will try to contact another farmer and tell him of the subsidy during the next time interval. To show how this is achieved in the model, we shall consider the five tellings which have to be made from cell (4, 4)($x_{4,4,0} = 5$). Figure 7.16 is a detailed map of the area surrounding cell (4, 4) which gives the location of farms within

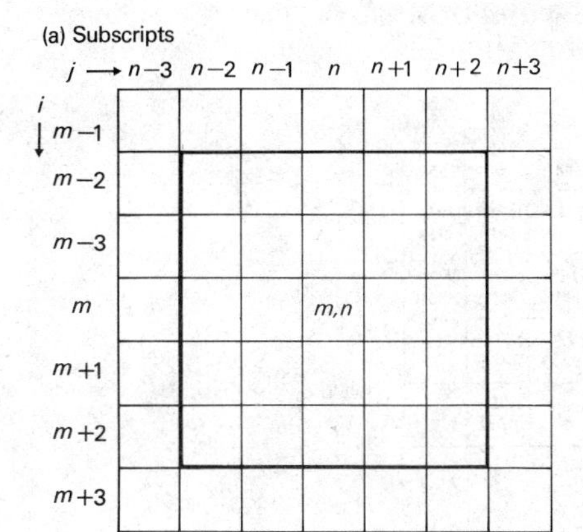

(b) Unweighted mean information field

$i \downarrow$ / $j \rightarrow$	$n-2$				$n+2$
$m-2$	·0096	·0140	·0160	·0140	·0096
	·0140	·0301	0547	·0301	·0140
$q_{mn}^{ij} =$	·0168	·0547	·4431	·0547	·0168
	·0140	·0301	·0547	·0301	·0140
$m+2$	·0096	·0140	·0168	·0140	·0096

(c) Number of farms per cell, (y_{ij})

$m = 4, \quad n = 4$

$i \downarrow$ / $j \rightarrow$	2	3	4	5	6
2	9	4	6	14	25
3	13	15	15	11	18
4	7	7	14	12	20
5	8	5	6	10	11
6	31	20	18	17	20

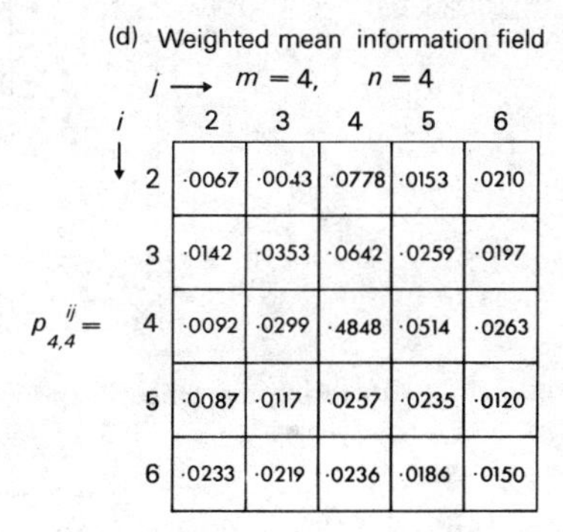
(d) Weighted mean information field

$m = 4, \quad n = 4$

	$i \downarrow$ / $j \rightarrow$	2	3	4	5	6
	2	·0067	·0043	·0778	·0153	·0210
	3	·0142	·0353	·0642	·0259	·0197
$p_{4,4}^{ij} =$	4	·0092	·0299	·4848	·0514	·0263
	5	·0087	·0117	·0257	·0235	·0120
	6	·0233	·0219	·0236	·0186	·0150

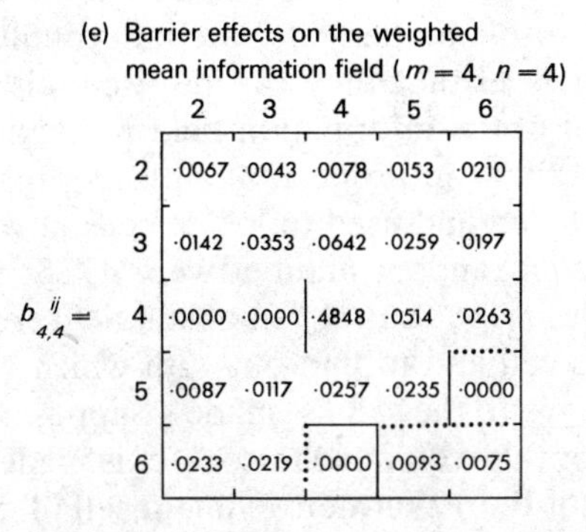
(e) Barrier effects on the weighted mean information field ($m = 4$, $n = 4$)

		2	3	4	5	6
	2	·0067	·0043	·0078	·0153	·0210
	3	·0142	·0353	·0642	·0259	·0197
$b_{4,4}^{ij} =$	4	·0000	·0000	·4848	·0514	·0263
	5	·0087	·0117	·0257	·0235	·0000
	6	·0233	·0219	·0000	·0093	·0075

FIG. 7.15. The structure of the weighted mean information field. Kindly provided by R. W. Thomas.

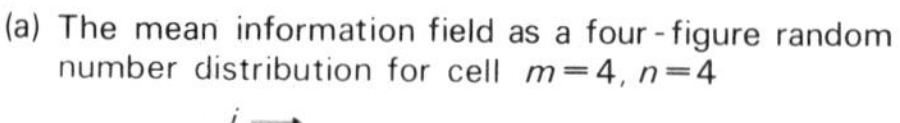

(a) The mean information field as a four-figure random number distribution for cell $m=4$, $n=4$

$i \downarrow$ / $j \rightarrow$	2	3	4	5	6
2	1 – 67	68 – 110	111– 168	168– 341	342– 551
3	552– 693	694– 1046	1047– 1688	1689– 1947	1948– 2144
4	×	×	2145– 6992	6993– 7506	7507– 7769
5	7770– 7856	7857– 7973	7974– 8230	8231– 8465	×
6	8466– 8698	8699– 8917	×	8918– 9010	9011– 9085

Wasted tellings = 9086–0000

(b) Five simulated contacts from cell 4,4 at $t=0$

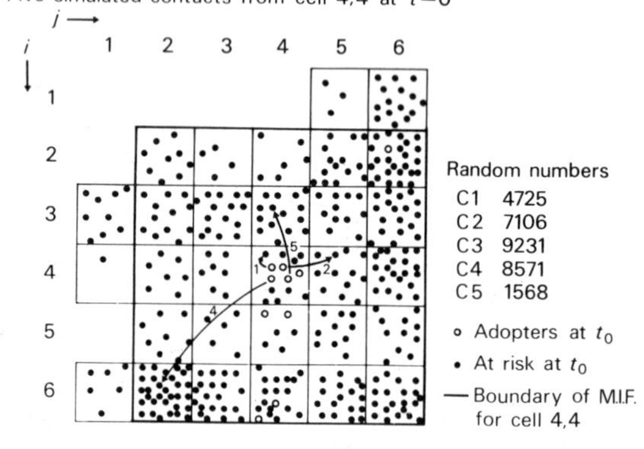

FIG. 7.16. Operating the Hägerstrand model. Kindly provided by R. W. Thomas.

each cell and the location of the farmers who had adopted the subsidy by 1929.

The unweighted mean information field is centred over cell (4, 4) and the set of weighted probabilities calculated. Secondly, the weighted probabilities are adjusted so that they can be used with four figure random numbers. Each cell in the information field is allocated a set of numbers between 0000 and 9999 in proportion with its probability. Five random numbers are then taken and used to locate cells in which the five contacts live. Suppose the first random number were 4725; this would mean that the contact is made in the central cell of the field (Figure 7.16). Some 13 farmers live in this cell and to find out with which one contact is made a random selection from the 13 is made. Suppose the second random number were 7106; this would mean contact is made between a farmer in cell (4, 4) and one of the 12 farmers living in cell (4, 5) a random selection from the 12 singling out the actual contact. This process is repeated until all farmers in cell (4, 4) who had already adopted the subsidy have made a

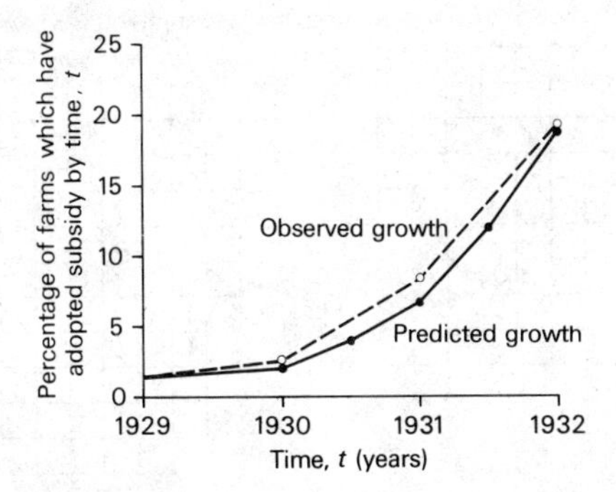

FIG. 7.17. The growth and timing of farm-subsidy diffusion. Kindly provided by R. W. Thomas.

contact. The mean information field is then centred on another cell which contains farmers who had adopted the subsidy in 1929, and so on until all the original 22 farmers who had adopted the subsidy have made a contact. The distribution of farmers who have adopted one subsidy by the end of the first time interval can now be mapped (Figure 7.14b). To find the pattern of adoption after the next time interval the entire process is repeated for all the farmers shown in Figure 7.14b and the resulting pattern of spread is shown in Figure 7.14c. And so the pattern of spread can be built up for a series of time intervals. It is usual in Monte Carlo simulations, because the pattern of spread in one run may vary considerably with the pattern of spread in another, to take the average pattern at each time step of several simulation runs.

The testing of the results of Monte Carlo models is usually a fairly informal procedure which entails matching predicted and observed patterns of spread. The difference between the observed and predicted growth of subsidy adoption between 1929 and 1932 is graphed in Figure 7.17 which shows that in all years the model slightly underpredicted the number of farmers who adopted the subsidy. And comparison of predicted and observed patterns (Figures 7.14 and 7.18) shows the model does capture the general diffusion of subsidy adoption, southwards and eastwards from the original sources in the vicinity of cell (4, 4).

7.5. Process–response models

The coupling of models of system structure and models of system process leads to process–response models. Using the terminology of Chorley and Kennedy (1971), as introduced as p. 9, a process–response system consists of conjoined morphological and cascading systems. The atmosphere, for example, may be thought of as a process–response system, the morphological components—air temperature, pressure, wind, humidity,

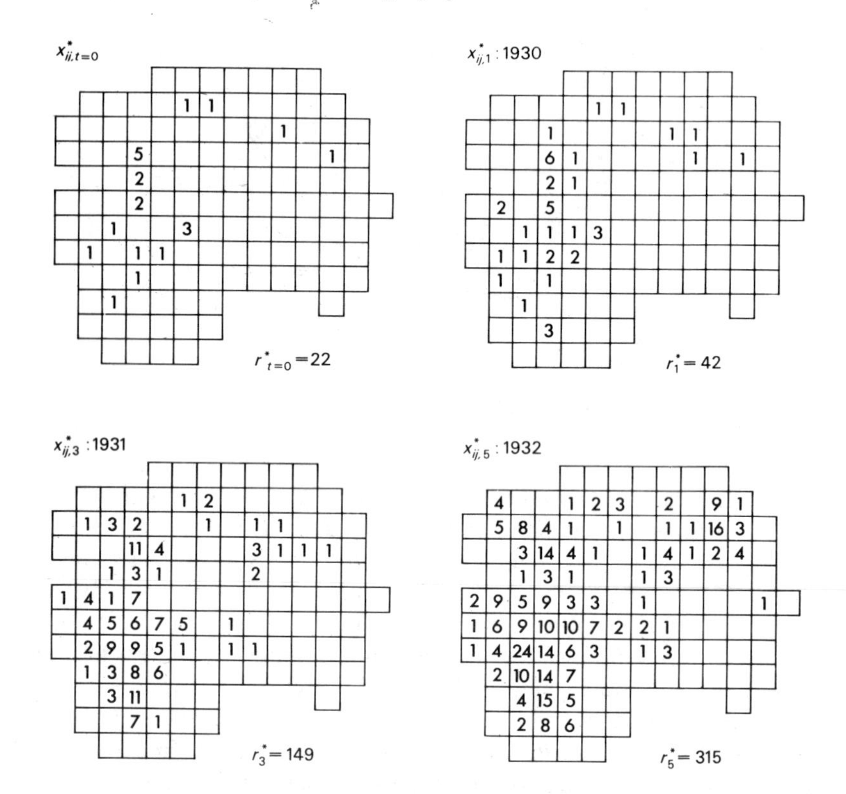

FIG. 7.18. Observed diffusion of farm-subsidy adoption in part of southern Sweden, 1929 to 1932. From T. Hägerstrand (1967), 'On Monte Carlo simulation of diffusion', in *Quantitative geography*, Part I: *Economic and cultural topics* (ed. W. L. Garrison and D. F. Marble), pp 1–32 (Northwestern Studies in Geography, No. 13). Northwestern University Press, Evanston, Illinois.

and turbidity—of which interlock with, and respond to, flows of energy, mass, and momentum (Terjung, 1976). In process–response models, the relation between form and process is usually a two-way interaction: process alters form and the changed form in turn modifies the process.

Analysis of process–response systems relies heavily on numerical-analytical modelling. The idea is, from a set of non-linear, partial differential equations which describe the dynamics of a process-response system, to predict the space–time variations is morphological system components. In the case of the atmosphere this may involve the prediction of air, surface, and soil temperature as well as moisture, wind, and pressure fields. In the case of hill-slopes this may entail predicting the form of a slope profile at different points in time. In both cases the model is a set of partial differential equations which are solved using either classical methods of integration or, in conjunction with a suitable spatial finite differencing scheme akin to the finite differencing scheme dealt with on p. 94, a numerical-integration technique. The models may be developed for one, two, or three dimensional problems. For instance, atmospherical models may consider space-time changes along a line, across an area, or within a volume. Few geographers are actively engaged in this kind of analysis but it is a growing field.

Not all process–response models are analysed by numerical procedures. Two other types of analysis are carried out. The first type, a statistical form of analysis favoured by many geographers, is applied when the laws governing the system process are ill known or where the rates of processes are unknown but surrogates for them are available. In a drainage basin, the laws which relate drainage density to inputs, storages, throughputs, and outputs of water are virtually non-existent. What is available are crude measures of flows—annual precipitation and evaporation, and run-off intensity, and factors which control flows—for instance, infiltration capacity, and the morphometric variable itself—drainage density. In this case analysis may entail the determination of statistical correlations between the variables. In a study of eighty drainage basins in the western USA, Melton (1957) found that potential evaporation, infiltration capacity, and run-off intensity account for 92.2 per cent of the variation in drainage density. Process–response models of this ilk are well described in Chorley and Kennedy (1971, chapter 4). Multiple-regression methods may also be used to establish relations between form and process variables. In coastal studies, the velocity of a breaking wave striking the coast at an angle is determined by flows such as the momentum of incoming water and form factors such as beach slope. Although a theoretical expression for velocity can be derived it is none too sound and empirical relations seem more useful at present. Harrison (1968), for example, found that

$$v = 0.170455 + 0.0373760C + 0.03180T + 0.241176H + 0.030923\beta$$

where v is the velocity, α the angle of wave approach, T the period of swell, H the significant breaker height, and β the beach slope.

The second type of analysis to which certain process–response systems may usefully be subjected is a version of phase plane analysis (p. 4). Process–response models of natural sedimentary systems seem particularly well suited to this type of analysis which we shall consider in more detail before looking at three examples of numerical-analytical models, two from physical geography—hill-slope development and global climatic models, and one from human geography—the spread of epidemics.

7.5.1. *Models of natural sedimentary systems*

Allen (1974, p. 226) has shown that natural sedimentary systems consist of surface forms which are roughly adjusted to a superjacent fluid or fluid-like agent. An example which springs to mind are dunes beneath the wind. The energy of the moving, fluid-like agent—the wind, is dissipated by the sedimentary system—the dunes, which will lead to a transference of sediment across the surface and hence to local changes in morphological variables such as dune heights and dune slope.

In modelling natural sedimentary systems it is usually possible to isolate just two attributes of interest, one which expresses the rate of energy supply to the system—the independent variable, and one which describes a morphological response—the dependent variable. The rate of energy supply is best stated as the velocity of the operative agent but fluid discharge or power supply per unit surface area may also be used. The descriptor of the morphological response may be the local erosion or deposition rate, the sediment load, the sediment concentration, or measures such as dune spacing or beach slope. The analysis of the system of interest then usually focuses on the dynamics of the form and process variables. Four dynamical properties are of special interest: equilibrium state, reaction time, relaxation time, and phase relations.

Equilibrium state. In sedimentary process–response models the notion of equilibrium generally means 'a tendency in the system for the magnitudes of the morphological attributes and the rate of energy supply to exist in a state of balance, at least statistically' (Allen, 1974). The equilibrium state may be identified for specific time scales because energy input—say incoming solar radiation—may exhibit cyclical behaviour about a mean state. Thus, although from year to year radiation receipt is more or less constant, there will be seasonal changes on which will be superimposed daily changes within which will be even smaller fluctuations arising from, say, changing cloud cover. In this example, seasonal changes are called first-order changes and have a period of about 10^7 seconds; diurnal changes are of the second-order and have a period of some 10^5 seconds; third order changes have a period of up to 10^3 seconds. Morphological attributes may respond differently to each order of energy input change.

Reaction time. It is common in sedimentary systems for the morphological response to trail behind the change in input. One reason for this time lag is the reaction time for a mechanism, or possibly several mechanisms, to respond to the changed input. The general pattern of response is shown in Figure 7.19a. In some cases, the reaction time is produced by threshold criteria for particle entrainment: for instance, a critical shear stress must be applied to a material in a river bed before

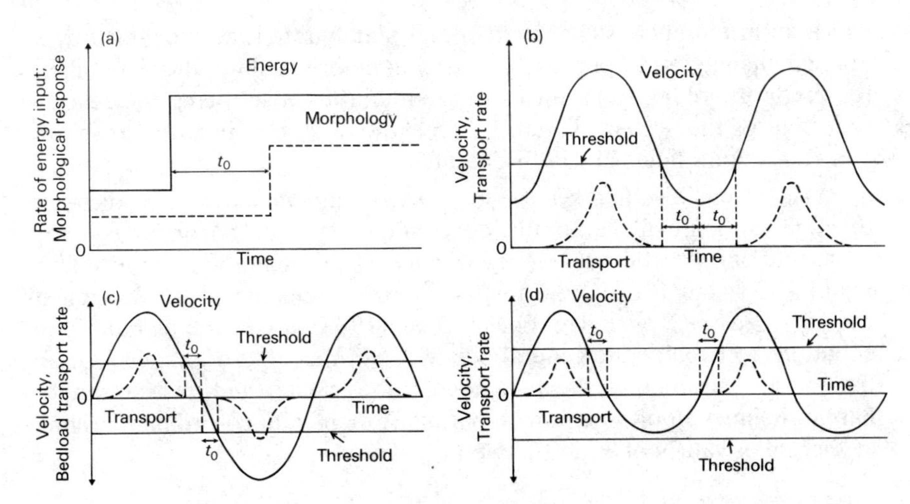

FIG. 7.19. The effect of reaction time on the morphological response in systems in which the rate of energy input is time-dependent. Reprinted with permission, from J. R. L. Allen (1974), 'Reaction, relaxation and lag in natural sedimentary systems: general principles, examples and lessons', *Earth-Science Reviews*, **10.** 263–342.

entrainment will take place and this phenomenon can lead to discontinuous transport if energy supply (river discharge) varies continuously with time (Figure 7.19b, c, and d). Another reason for the existence of a reaction time is the spatial separation of the energy input and the morphological response: pyroclastic material ejected from a volcanic vent cannot bring about a morphological response (by dint of changing the elevation of the sedimentary surface in the vicinity of a volcano) until it has travelled through the atmosphere.

Relaxation time. Relaxation time occurs in sedimentary systems when the morphological change cannot keep pace with a change in energy supply rate, a new state of system equilibrium being delayed independently of any reaction-time effects. Figure 7.20 as shows the situation in

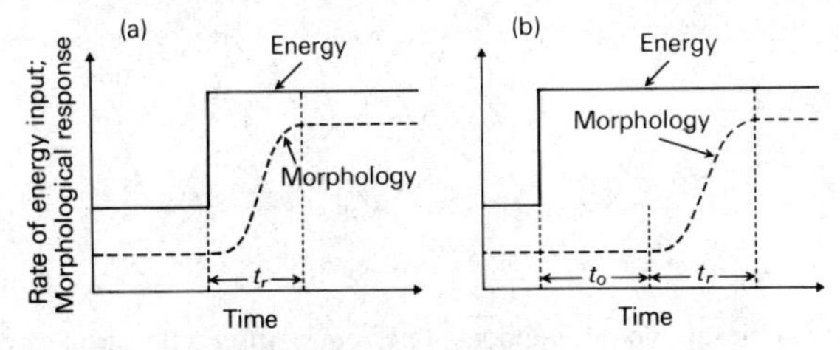

FIG. 7.20. The effect of (a) relaxation, and (b) reaction plus relaxation in systems in which the rate of energy input is time-dependent. Reprinted with permission, from J. R. L. Allen, op. cit Figure 7.19 above.

which an instantaneous change in energy supply rate is accompanied by a gradual change in morphology to a new equilibrium state, the time taken for the morphology to attain its new equilibrium level being the relaxation time of the system. Figure 7.20b shows a system in which reaction and relaxation effects are both operative.

Phase relations. In a system in which inputs vary cyclically through time, relaxation will cause the concomitant cyclical morphological response to lag behind the energy change. The situation is depicted in Figure 7.21; the two curves are out of phase and the phase difference can be measured as a lag time. Allen (1974) used simple harmonic equations to explore the general behaviour of this kind of system. If z is the energy variable, y the morphological variable, and z is a simple harmonic function of time, t, with period T, then, with co-ordinate origins at the mean values of y and z, we may write

$$y = a_y \sin\left(\frac{2\pi t}{T} - \alpha_y\right)$$

$$z = a_z \sin\left(\frac{2\pi t}{T} - \alpha_z\right)$$

where a_y and a_z are the amplitudes of y and z, and α_y and α_z are their phase angles; the quantity $(\alpha_y - \alpha_z) = \delta$ is called the radian phase difference. By expanding these equations and adding them, time may be eliminated and the relation between y and z found to be of the form

$$\frac{y^2}{a_y^2} + \frac{z^2}{a_z^2} - \frac{2yz}{a_y a_z} \cos \delta - \sin^2 \delta = 0.$$

This equation describes an ellipse, or if $\delta = 0$, a straight line, and if $\delta = \pi/2$ or $\pi/3$ a circle (Figure 7.22). The phase relationships between y

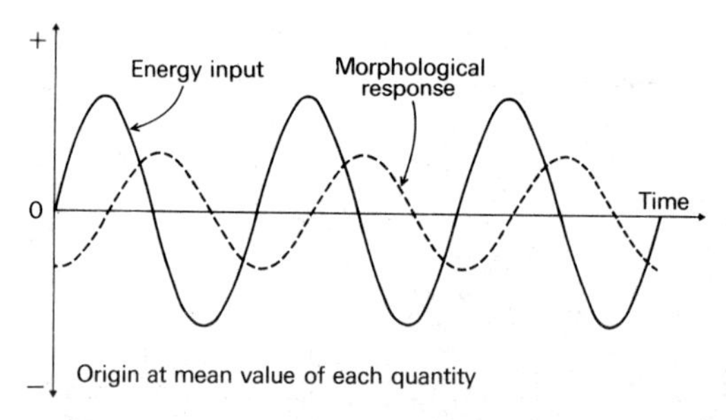

FIG. 7.21. Two simple-harmonic functions, one representing rate of energy input and the other morphological response, of the same period but different phase and amplitude. Reprinted with permission, from J. R. L. Allen, op. cit. Figure 19 above.

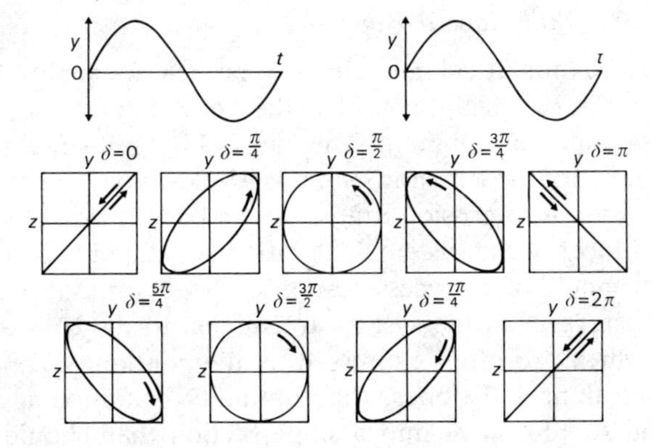

FIG. 7.22. Phase diagrams over a range of phase differences representing the variation of a quantity y with a quantity z when both are simple-harmonic with respect to time t. Reprinted with permission, from J. R. L. Allen, op. cit. Figure 7.19 above.

and z are not very appropriate for actual sedimentary systems. However, by using more realistic relations between y and z, phase diagrams may be prepared which can be used for assessing phase diagrams of actual systems, such as that produced by plotting suspended sediment rate against stream discharge (Figure 7.23).

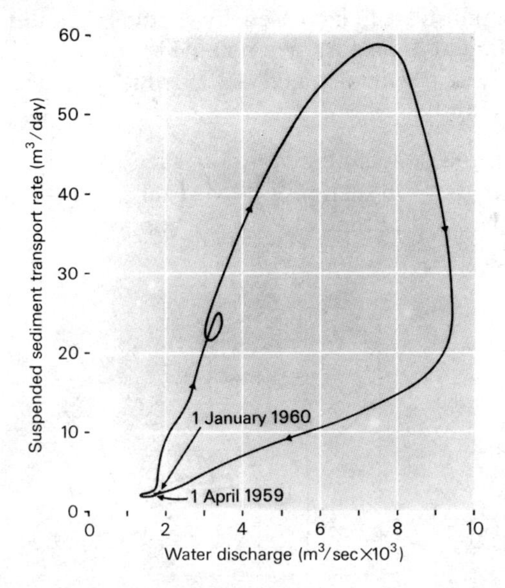

FIG. 7.23. Variation in the total suspended sediment transport rate with water discharge in the Forcados River, Onya, Nigeria, between 1 Apr. 1959 and 1 Jan. 1960. Reprinted with permission, from J. R. L. Allen, op. cit. Figure 7.19 above.

7.5.2. *Models of hill-slope development*

To illustrate the numerical modelling approach to analysing process–response systems, we shall consider the formulation and analysis of models of hill-slope development. The idea behind hill-slope models is to produce an equation which shows how land elevation, the form facet of the system, responds to erosion, transport, and deposition of sediment on it, the process facet of the system. The rationale behind hill-slope models, and any other model of a process–response system for that matter, is that the flows and storages in the system are subject to the laws of conservation (p. 93) which are often expressed as the continuity condition. For debris moving along hill-slopes, the continuity condition means that if more material should move into a slope section than should move out, then the difference will represent aggradation in the section; if less material should move into a slope section than should move out, then the difference will represent degradation in the section; if inputs and outputs should balance, the slope section will remain unchanged (Kirkby, 1971). We are now in a position to derive a hill-slope model.

A slope profile of horizontal extent X is divided into small sections of length Δx (Figure 7.24). The variation in height, h, with distance at time t is shown by the lower line in Figure 7.24. Owing to sediment transport over the slope during a time interval, Δt, the slope profile may be modified. At point x, the height of the land surface will have decreased or, in the case illustrated, increased by a small amount, Δh. The mass balance of sediment transport per unit width of slope in the section of slope lying between points x and $x+1$, which we shall call section x (Figure 7.24), may be written

$$\begin{array}{l} \text{sediment stored} \\ \text{in section } x \\ \text{at time } t+1 \end{array} = \begin{array}{l} \text{sediment stored} \\ \text{in section } x \\ \text{at time } t \end{array} - \left(\begin{array}{l} \text{sediment} \\ \text{inputs at} \\ \text{point } x \end{array} - \begin{array}{l} \text{sediment} \\ \text{outputs at} \\ \text{point } x+1 \end{array} \right) \Delta t$$

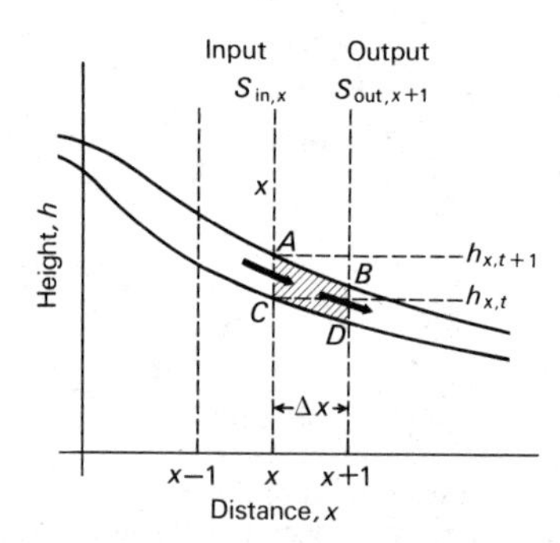

FIG. 7.24. The continuity equation applied to sediment movement on a hill-slope.

where the term in brackets, the inputs minus outputs, is a state transition function. The balance for the same section of slope can also be written in terms of the height changes

$$\begin{array}{l}\text{sediment stored}\\ \text{in section } x\\ \text{at time } t+1\end{array} = \begin{array}{l}\text{sediment stored}\\ \text{in section } x\\ \text{at time } t\end{array} + \left(\begin{array}{l}\text{height of}\\ \text{section at}\\ \text{time } t+1\end{array} - \begin{array}{l}\text{height of}\\ \text{section at}\\ \text{time } t\end{array}\right)\Delta x$$

which simply states that the change in sediment in the slope section is the difference between the volume of sediment in the section at time $t+1$ and the volume of sediment in the section at time t.

We now have two expressions which define the sediment stored in the slope section at time $t+1$. Using the following symbols

$$\begin{aligned} S_{x,t} &= \text{sediment stored in section } x \text{ at time } t\\ S_{x,t+1} &= \text{sediment stored in section } x \text{ at time } t+1\\ h_{x,t} &= \text{height of section } x \text{ at time } t\\ h_{x,t+1} &= \text{height of section } x \text{ at time } t+1\\ S_{\text{in},x} &= \text{sediment inputs at point } x\\ S_{\text{out},x+1} &= \text{sediment outputs at point } x+1 \end{aligned}$$

the two expressions may be equated to give

$$-(S_{\text{in},x} - S_{\text{out},x+1})\,\Delta t = (h_{x,t+1} - h_{x,t})\,\Delta x$$

which tells us that the change in height in the slope section depends on sediment inputs and outputs. The term $(S_{\text{in},x} - S_{\text{out},x+1})$ may be written as ΔS so the equation then looks like this

$$-\Delta S\,\Delta t = (h_{x,t+1} - h_{x,t})\,\Delta x$$

or, after rearrangement,

$$h_{x,t+1} = h_{x,t} - \left(\frac{\Delta S}{\Delta x}\right)\Delta t. \qquad (7.24)$$

This very useful equation enables us to work out how the height of the slope section changes in response to sediment transport processes. To make the equation operational, the state transition function $\Delta S/\Delta x$ needs to be defined by a sediment transport law. A common supposition is that sediment transport at any point on a slope profile depends on the slope gradient at the point and some function of the distance of the point from the watershed (this is roughly the distance of overland flow). With the following designations

$f(x)$ = some function of distance from watershed, and

$\frac{\Delta h}{\Delta x}$ = slope gradient

we may write the sediment transport law as

$$S = f(x)\left(-\frac{\Delta h}{\Delta x}\right)$$

and so

$$\Delta S = \Delta\left\{f(x)\left(-\frac{\Delta h}{\Delta x}\right)\right\}.$$

Putting this definition of ΔS into equation 7.24 produces

$$h_{x,t+1} = h_{x,t} - \left[\frac{\Delta}{\Delta x}\left\{f(x)\left(-\frac{\Delta h}{\Delta x}\right)\right\}\right]\Delta t. \tag{7.25}$$

Equation 7.25 is more or less the final model and it relates the change in height of the slope section to morphological properties of the slope. To use the model, the function $f(x)$ needs specifying. Empirical work suggests that $f(x) = x^m$ where m varies according to the sediment-moving process operative on the slope: for soil creep $m = 0$ and for slope wash $m = 1$. Hirano (1975) adds two other parameters, a and b, to this function to give

$$f(x) = a + bx^m.$$

And putting this expression into equation 7.25 produces

$$h_{x,t+1} = h_{x,t} - \left[\frac{\Delta}{\Delta x}\left\{(a - bx^m)\left(-\frac{\Delta h}{\Delta x}\right)\right\}\right]\Delta t. \tag{7.26}$$

Expanding all the brackets in the state transition function (see Wilson and Kirkby, 1975, p. 137 for details of how to do this) we get

$$h_{x,t+1} = h_{x,t} + \left[a\frac{\Delta^2 h}{\Delta x^2} + bx^m\frac{\Delta^2 h}{\Delta x^2} + b\frac{\Delta x^m}{\Delta x}\frac{\Delta h}{\Delta x}\right]\Delta t. \tag{7.27}$$

Hirano (1975) simplified this equation to

$$h_{x,t+1} = h_{x,t} + \left[a\frac{\Delta^2 h}{\Delta x^2} + b\frac{\Delta h}{\Delta x}\right]\Delta t. \tag{7.28}$$

Equation 7.28 is the finished model; it states that the height at any point on a slope profile changes in proportion to the slope curvature at the point, $\Delta^2 h/\Delta x^2$, and in proportion to the slope gradient at the point, $\Delta h/\Delta x$. The parameter a, which is called a subduing coefficient, determines the importance of slope curvature in bringing about a height change; the parameter b, or recessional coefficient, determines the importance of slope gradient in bringing about a change of height. To see how the model (equation 7.28) operates in a given case, three things must be specified: the values of parameters a and b; the initial form of the slope profile; and the height at top and bottom of the profile during the time period of interest—these are the boundary conditions. Hirano (1975) considered several interesting cases. In one case the parameters were set at $a = 0.25$ and $b = 2.0$; the initial slope profile was a horizontal line and might represent a plateau surface; the boundary condition at one end of the slope represented down-cutting by a river and the boundary condition at the other end represented a watershed so that $\Delta h/\Delta x$ at this point was set equal to zero. In fact, three different types of down-cutting were considered: the case in which down-cutting proceeds at an accelerating rate; the case in which down-cutting proceeds at a decreasing rate; the case in which down-cutting proceeds to accelerate at first, then stays constant for a while and later gradually decelerates. The predicted changes in slope form through time using the model are shown in Figure 7.25. The first two patterns of slope change, corresponding to the first two

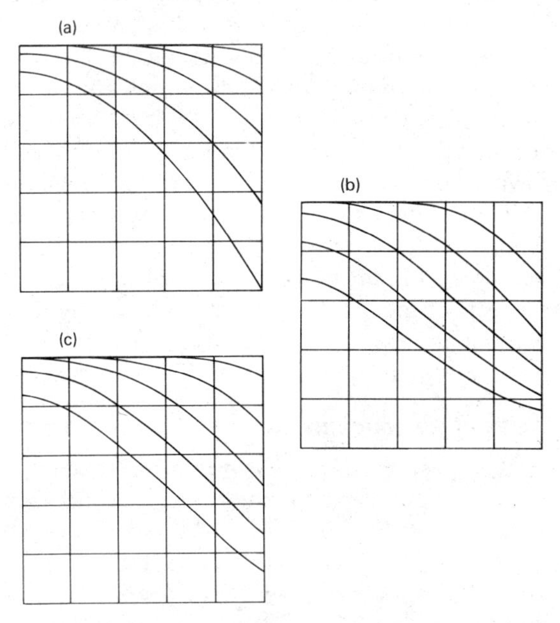

Fig. 7.25. Predicted patterns of slope change under varying assumptions of downcutting rate: (a) accelerating rate (b) decreasing rate; and (c) accelerates at first, is then held constant for a while, and later decelerates. Reprinted with permission from M. Hirano (1975), 'Simulation of development process of interfluvial slopes with reference to graded form', *Journal of Geology*, **83.** 113–23, Copyright 1975. The University of Chicago.

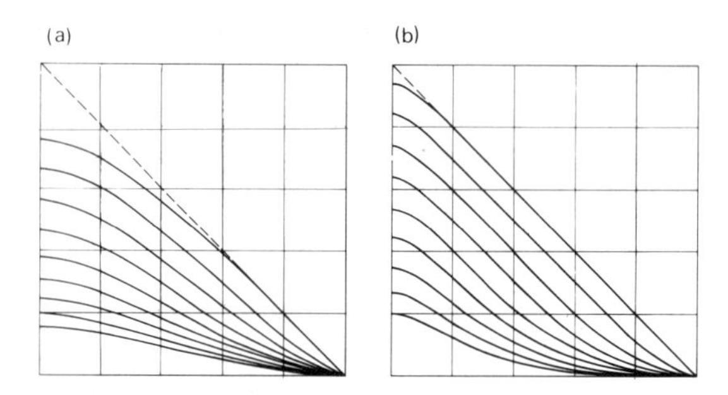

FIG. 7.26. The effect of the parameters a and b on slope development. Reprinted with permission, from M. Hirano, op. cit. Figure 25 above. Copyright 1975. The University of Chicago.

cases of down-cutting, describe Penck's 'aufsteigende Entwickling' and 'absteigende Entwicklung' respectively. In all cases, an equilibrium slope form is never attained. Hirano (1975) found in other experiments that a steady state slope form is established only if the rate of river down-cutting is constant.

The significance of the a and b parameters in the model is demonstrated in Figure 7.26. In the two cases shown, the initial form of the slope profile is a straight slope, height at the valley bottom is held constant, and the boundary condition at the watershed is such that $\Delta h/\Delta x = 0$ at all times. The ratio of a/b is 0.25 in case one and 0.0625 in case two: the effect of this is that in the first case the slope profile tends to wear down whereas in the second case the slope profile tends to wear back. This is interesting because the parameter determines the relative effect of the $\Delta^2 h/\Delta x^2$ term in the model which probably represents soil creep and creep-like processes whereas the b parameter determines the relative effects of the $\Delta h/\Delta x$ term which probably represents surface wash and wash-like processes. It would appear that where soil creep is the main process operative on the slope, wearing back takes place.

7.5.3. *Global-climatic models*

The most sophisticated process–response models have been built by atmospherical physicists and meteorologists to simulate the general atmospherical circulation. The workings of these models are exceedingly complicated. Nonetheless, the value of global–climatic models can be appreciated by studying an example. One of the simpler models was developed by Sellers (1969). Based on the thermodynamic equation for the earth-atmosphere system, Seller's (1969) model used only annual average conditions and made a number of gross assumptions in relating all variables to mean annual sea-level temperature distribution; it yielded many interesting results. For instance, the removal of the north-polar ice would, according to the model, increase the temperature polewards of 70°N by a maximum of 7°C, tropical temperatures by 1°C, and temperatures near the South Pole by between 1 to 3°C. Removal of the Antarctic ice would produce a rise of temperature in the Antarctic of 12 to 15°C and a 4°C rise in the Arctic. Removal of ice from both polar regions would yield a 7 to 10°C rise in the Arctic and a 13 to 17°C rise in the Antarctic; warming in the tropics would never exceed 2°C. The effects of changing the solar constant on climate were also investigated with the model. It would seem that a 2 per cent decrease in the solar constant is sufficient to cause another ice age with ice caps extending equatorwards to 50° and mountain glaciers and heavy winter snow to 30°. Further reduction of the solar constant would lead to a rapid transition to an ice-covered Earth with an equilibrium temperature of −100°C; but, as Sellers (1969) pointed out, the assumptions of the model make predictions of such extreme conditions highly speculative. A 3 per cent increase in the solar constant would be sufficient to melt the existing ice caps. Averaged over the Earth's surface, the energy likely to be consumed by man and converted into heat two centuries hence will be little more than 5 per cent of the solar radiation intercepted by the globe each year; the effect of this on climate however would be considerable: globally the temperature would rise an average of 15°C ranging from 11°C near the equator to 27°C at the North Pole. Thus, according to the model, if man's energy-consuming activities continue unabated, the ice caps would eventually be eliminated and a climate much warmer than that of today would develop.

In a later paper, Sellers (1973) developed another global–climatic model which allowed for seasonal coupling and interaction between continents and oceans. Using present-day values for input parameters, the model reproduces well many of the chief thermal and dynamical characteristics of the Earth–atmosphere system. Among the more interesting results from the model is the finding that the Northern Hemisphere is extremely sensitive to factors which would produce climatic change. For example, a 0.6 per cent decrease of the solar constant would lead to a 3.5°C drop in temperature at 65°N in February and 4.1° C in August; the corresponding temperature drops for 65°S are 1.1°C and 1.7°C and for the globe as a whole 1.6°C and 1.8°C.

Many models of this type are being built; some recent examples may be found in North (1975) and Pollard (1978).

7.5.4. *The spread of the Black Death*

An interesting field of human geography in which process–response models have shed some light is the study of the spread of epidemics.

Noble (1974) set up a numerical process–response-type model of the geographical spread of epidemics on the assumptions that:

(1) There are just two interacting populations—the infectives and the susceptibles.
(2) The change in the infective population, I, in a small area over a small period of time is equal to the rate of transitions from the susceptible population (that is the number of susceptibles in the small area per unit time which became infected), less the removal rate of infectives, which is equal to the death-rate of the infectives, plus the recovery rate, plus the net migration from the area.
(3) The change in the susceptible population, S, in a small area over a small period of time can be expressed in a similar way to the change in the infective population, except that transitions from susceptible to infective population (owing to a susceptible contracting the disease) appear as a net loss and the rate of removal is simply the net migration.
(4) Under conditions prevailing during plagues, the net migration of infective and susceptible populations result from random walks, as opposed to mass migrations, and can thus be represented as a simple diffusion process.

With these assumptions, the mathematical model may be formulated as

$$\begin{matrix}\text{infective}\\ \text{population}\\ \text{at point } x, y\\ \text{at time } t+1\end{matrix} = \begin{matrix}\text{infective}\\ \text{population}\\ \text{at point } x, y\\ \text{at time } t\end{matrix} + \left(\begin{matrix}\text{transition rate}\\ \text{of susceptibles}\\ \text{to infectives}\end{matrix} - \begin{matrix}\text{deaths of}\\ \text{infectives}\end{matrix} + \begin{matrix}\text{net migration}\\ \text{of infectives}\end{matrix}\right)\Delta t$$

$$\begin{matrix}\text{susceptible}\\ \text{population}\\ \text{at point } x, y\\ \text{at time } t+1\end{matrix} = \begin{matrix}\text{susceptible}\\ \text{population}\\ \text{at point } x, y\\ \text{at time } t\end{matrix} + \left(-\begin{matrix}\text{transition rate}\\ \text{of susceptibles}\\ \text{to infectives}\end{matrix} + \begin{matrix}\text{net migration}\\ \text{of susceptibles}\end{matrix}\right)\Delta t.$$

Expressing these ideas in symbols, with a state transition function in curly brackets containing spatial terms, we may write

$$\begin{aligned} I_{x,y,t+1} &= I_{x,y,t} + \left\{KIS - \mu I + D\left(\frac{\Delta^2 I}{\Delta x^2} + \frac{\Delta^2 I}{\Delta y^2}\right)\right\}\Delta t \\ S_{x,y,t+1} &= S_{x,y,t} + \left\{-KIS + D\left(\frac{\Delta^2 S}{\Delta x^2} + \frac{\Delta^2 S}{\Delta y^2}\right)\right\}\Delta t \end{aligned} \qquad (7.29)$$

where S and I are the susceptible and infective population densities respectively; μ is the infectives' death-rate; K is a transmissibility coefficient and measures the average area of infection swept out by an infective per unit time; D is a diffusion coefficient; and the bracketed partial-difference terms represent the random net migrations. The initial population density of susceptibles is called U. Noble (1974) calibrated the model (equations 7.29) for the spread of the Black Death of 1347–50 in Europe for a one-dimensional case. The initial conditions and parameters were: population density $U = 50$ per square mile (estimate for Europe in AD 1347); diffusion coefficient 10^4 square miles per year (based on minor news or gossip spreading at 100 miles per year); the transmission coefficient 0.4 square miles per year (the area swept out in walking at a 1 mile per hour slow walk assuming a 5 foot flea hop and a 10 per cent transmission probability); mortality rate of infectives 15 per year. With these parameters the velocity of spread ($v = \sqrt{KUD}$) is 450 miles per year which agrees well with the observed rate of spread. Figure 7.27 shows the model solved with these parameters for a one dimensional case. The wave-like progress of the infective population away from the region of the original infection is evident. Noble showed that for the plague to spread in this manner the ratio μ/KU, known as the damping parameter,

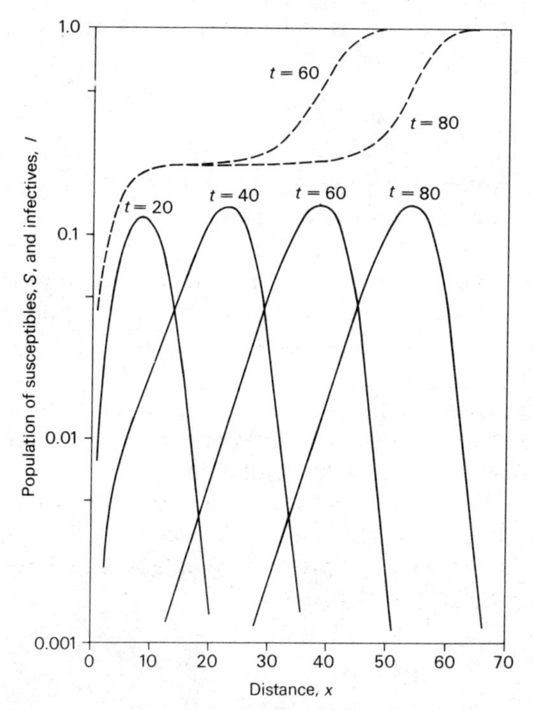

FIG. 7.27. One-dimensional spread of the Black Death. Reprinted with permission, from J. V. Noble (1974), 'Geographic and temporal development of plagues', *Nature*, **250** 30 Aug. 1974), 726–9.

Table 7.11
Parameters of plague waves for several values of the damping parameter

Damping parameter (μ/KU)	Velocity of propogation (in units of $\sqrt{KUD}$)	Maximum infective population density	Fraction of survivors
0.10	2.0	0.7	0
0.25	1.9	0.4	0.025
0.50	1.5	0.14	0.22
0.75	0.8	0.03	0.6
0.90	0.55	0.004	0.8

Reprinted with permission, from J. V. Noble (1974), 'Geographic and temporal development of plagues', *Nature*, **250** (30 Aug. 1974), 726–9.

must be less than unity. This can be explained in physical terms since if the death-rate, μ, is high, the infectives will die too quickly to spread the disease; and if the population density, U, times the transmissibility coefficient is too small, the plague cannot be transmitted. It can also be shown that the minimum or critical population density, U_{crit}, necessary for an outbreak of the plague is

$$U_{crit} = \mu/K.$$

In this, example U_{crit} is 37.5 per square mile. Some of the plague wave parameters—the velocity of propagation, the maximum infective population density, and the fraction of survivors—are related to the damping parameter μ/KU as is shown in Table 7.11. Paradoxically, the more rapidly fatal the disease is for an individual, the better it is for the population as a whole. Thus, the survivors of high-mortality plagues ($\mu/KU > 0.5$) are either immune or lucky but the actual fraction of the whole population is large.

The model is highly simplified—in reality, population density is not uniform but clustered for instance—but has given some interesting results which, in the case of the Black Death, match observed patterns quite well. Also the mathematical analysis of the model has revealed several interesting measures of the spread of epidemics and the relations between them. Clearly, the model might prove profitable in studying other particular epidemics of geographical phenomena.

8 Prospect

> One by one exceptions vanish,
> and all becomes systematic.
> SPENCER, *Social Statics*

THE reader should now have some idea of just what systems analysis is capable of. In this the last chapter of the book, we shall, in the first place, discuss a special quality of systems analysis—its ability to model several interconnected system components acting in concert; in the second, tell of the advances made in modelling spatial systems; and, to conclude, describe the problem of modelling system evolution.

8.1. Functional connections

The viewing of connections between system components as functional relations is an innovation in geography; it is, perhaps, best appreciated in the light of methodological changes which from time to time have overturned the canons of geographical enquiry. These overturnings, and there have been at least four of them, have gone hand in hand with the progress of geography, each having begun a new stage in geographical thinking. The first stage involved field study of systems of interest—soils, landforms, plants, settlements, and so on—to establish criteria for their classification; in other words, sorting out morphological system components. In the second stage, interest was directed to describing the development of morphological system components as a progression in time in response to some underlying genetic process. The work of W. M. Davis on landform development is a case in point. Today, research at this stage is still pursued by many workers and usually entails the establishing of correlations, or more sophisticated statistical measures of association between system components; it is hampered by the problem of defining system boundaries, the large number of variables needed, and the gigantic sampling designs required (Terjung, 1976). The aim of such analysis is to elucidate cause and effect mechanisms purely by examining the morphology of a system and its parts. The third and fourth stages both entail seeing how systems function, how their components are related, and how the relations bring about changes of system state. In the third stage, the passage of energy, material, and information through systems is considered without referring to the spatial distribution of the energy, mass, or information transactions; this stage then is the study of flow systems as outlined in chapter 6 and is a very active research area. Research in the fourth stage considers the pattern of energy, mass, or information transactions within some spatial domain—locational facets of systems are brought into the analysis; this stage is thus the study of regional systems as discussed in chapter 7, and, like the studying of flow systems, is an active and new area of research.

At least two immensely important advantages accrue from building functional models of systems. For one, new pedagogical insights into a subject may be gained. Consider climate; Terjung (1976) has said that traditionally this is a long-term, spatial generalization of day-to-day meteorological conditions or, more accurately, morphological components of the atmosphere. Climatology based on traditional concepts is highly descriptive and contains rudimentary explanations. Climate is better thought of as a process–response system produced by a two-way link between on the one hand cascades of energy, mass, and momentum and, on the other, morphological components of the atmosphere and the configuration of the Earth's surface; the climate of a particular region is simply the state or states of the system which seem to be typical of that area. Similarly, in biogeography, Gersmehl (1976) claimed his three-compartment systems model of mineral circulation in ecosystems (see p. 145) has two valuable functions for the teacher of geography. First of all, it highlights the connectedness of the natural environment and emphasizes the folly of assuming simple cause-and-effect links between pairs of system components. In the second place, the model embodies the idea of a zonal ecosystem as a regionally unique combination of mineral-flow laws which imposes a powerful set of constraints upon human activity. And in human geography, Reif (1973) has indicated the benefit of regarding a city or a region as a system; his argument runs as follows. The parts of a city or region cannot be studies in isolation because they are strongly interconnected. So the location of a shopping centre cannot be studied without taking account of the number of likely customers and transport facilities, a change in any one of the three factors inducing a change in the other two. In turn, these three factors are tied up with economic factors, such as changes in income level, which can alter the size of the shopping centre, the mobility of shoppers, and the residential location of people. In short, all the factors are best considered to act together as a functional system.

The second important advantage accruing from a functional view of systems is the feasibility of studying systems which are too large, too dangerous, or otherwise impossible to experiment with directly in the field. We have seen an example of this in global-climatic models. It is clearly impractical to determine the long-term effects of a sustained climatic change by direct experiment. Computer simulation using a systems model has proved an effective surrogate for doing this. In a study by

Cooper *et al.* (1974), five ecosystem models were operated in parallel under perturbations of climate. The aim was not to produce precise predictions of climatically induced ecological change, but to see if qualitative patterns could be discerned in ecosystem response to climatic stress. Several trends emerged which could not necessarily have been predicted by traditional methods of ecological analysis. Primary plant production decreases linearly with temperature decrease, except in extremely arid regions and in semi-arid areas with coniferous trees where it increases; the greatest decreases occur in ecosystems well supplied with water. Consumer organisms are non-linearly affected by climatic perturbations in a more severe and less predictable manner than are primary producers: in a tundra ecosystem, the simulation model predicted an entirely new structural relationship among the system components—primary producers, consumers, and predators—as a consequence of a modest climatic change. Rates, pathways, and the seasonal course of microbial decomposition are altered in complex ways by climatic change, the changes being likely to alter long-term successional responses. Depending partly on the initial climate and the successional status of the community being simulated, total ecosystem standing crop may be increased or decreased by climatic change. Water stress in plants is reduced slightly by a decrease in temperature, even if the temperature decrease goes hand in hand with a reduction in rainfall. Importantly, the effects of climatic forcing functions considered together in the model are different from the effects produced in models in which climatic factors are varied one at a time.

Forrester's (1969) model of an urban system focuses on a set of interacting businesses, houses, and people (see p. 134) and describes the growth of a city over 250 years starting from a nearly empty land area; it can also be used to test the effect of urban policies before they are put into practice. An example is given by Alfeld and Meadows (1974) who noted the present problem of deterioration and rising prices of housing in American cities, and the reaction to this state of affairs—the construction of low-cost housing. But, they wondered, bearing in mind the complex repercussions that the building of low-cost housing might have throughout the urban system, will a low-cost housing programme alleviate a city's housing problem? Computer simulation of urban change in response to a 5 per cent annual addition of housing stock shows that as the availability of under-employed (low-cost) housing increases, so more under-employed workers are attracted to the city and the total under-employed worker population rises for a decade. The giving over of available land to low-cost housing raises land prices and so depresses the establishment of new industry. Increased under-employed workers and fewer new job opportunities act, in the long run, to increase local unemployment rates and so retard the large immigration of under-employed to the city. And, because of increased local population densities, housing conditions in the city have not materially improved after fifty years, by which time a new steady state has become established that is worse than that which prevailed at the inauguration of the low-cost housing programme.

8.2. Spatial connections

The success of systems analysis in geography may ultimately depend on its ability to crack spatial problems. The regional models described in chapter 7 are but a small sample of studies which incorporate locational factors. For instance, though a family of interaction models were briefly considered, many other spatial models have been built to help solve problems of urban and regional development. As these are, or should be, of special interest to geographers, we shall examine some of them in a little detail.

There are two, complementary ways of modelling urban form (Harris, 1972). The first way, static modelling, leads to a steady-state description of an urban system without considering the dynamics of the processes which fashion the steady state. The other way, dynamic modelling, pays attention to the dynamic processes which mould the urban form, may or may not lead to a steady-state solution, and provides a description of urban (or regional) growth. Included in the static modelling group is the famous model of a metropolis, built by Lowry (1964), which defined the steady-state distribution of population and service activities for Pittsburgh based upon the transport system, travel patterns, and the location of basic industry. Another example is the land-use-plan design model of Schlager (1965) which included three sets of variables—type of land use (quality variables), density of land use (quantity variables), and geographical location (locational variables). The variables were combined as a so-called cost function which is subject to a set of constraints or restrictions such as demand for each kind of land use, limits on land use in each zone, and the proportion of one land use allowed relative to another land use in the same or a different zone. Using a procedure called linear programming, the cost function was minimized to find the spatial combination of land-use type and density within a city which is, mainly in economic terms, cheapest to develop.

Dynamic models include a growth-allocation model for the Boston region, which covered all aspects of urban location, constructed by Hill (1965). This model considered the highly interrelated nature of locational preferences of the population, employment, and other components of the urban system, as well as exogenous factors that influence the development of urban land. Spatially detailed patterns of urban activities were predicted some twenty years into the future in the light of existing activities, externally forecast regional growth, and exogenous policies concerning the development of transport, open space, zoning, public services, and regional growth.

White (1977) set up a dynamic interaction model to study the growth of central place systems. In essence, he represented each activity in each centre in a central place system by a difference equation of the form

$$\begin{matrix}\text{size at}\\ \text{centre } i \text{ of}\\ \text{activity } j\\ \text{at time } t+1\end{matrix} = \begin{matrix}\text{size at}\\ \text{centre } i \text{ of}\\ \text{activity } j\\ \text{at time } t\end{matrix} + \begin{matrix}\text{growth}\\ \text{factor}\end{matrix}\left(\begin{matrix}\text{revenue}\\ \text{function for}\\ \text{activity } j\\ \text{in centre } i\\ \text{at time } t\end{matrix} - \begin{matrix}\text{cost}\\ \text{function for}\\ \text{activity } j\\ \text{in centre}\\ \text{at time } t\end{matrix}\right).$$

The assumption here is that the growth (or decline) of each centre depends on its profitability—when a centre attracts more revenue than is used in providing goods and services, it will grow; when costs exceed revenue, decline will set in. The cost function included fixed and marginal costs for each activity at each centre. The revenue function was calculated according to a gravity equation or an exponential interaction equation. In White's examples, the landscape was divided into graid cells, among which population was evenly spread except in those cells containing central places. All grid cells were potential contributors to a centre, but the gravity and exponential interaction assumptions incorporate a distance–decay effect such that nearer cells will contribute more revenue than cells farther away. An important parameter in the revenue function is the distance–decay exponent (see p. 158) because this determines how sharply the revenue from a cell drops off with increasing distance from a centre.

The model was run for a central place system of, usually, up to 20 centres and two activities on a map of 2500 grid cells. Three different starting sizes of centre were used: all centres the same size, large centres only, and small centres only. Several different map layouts were used including a standard map of 20 centres located randomly on a 50×50 grid, and a regular map of 25 centres arranged on a 60×60 grid to form a regular rectangular pattern. The results of the model revealed the special significance of the distance–decay exponent in the revenue function. In the gravity equation, values for the exponent in the range 0.5 to 3.0 were tried because this is the range of values found in empirical studies of consumer behaviour. It was found that the effect of the distance-decay exponent is largely independent of initial conditions. In general, for the higher exponents, distance to neighbouring centres once a steady state is established is the main determinant of the size of centres; for the lower exponents, the centrality of a centre in the entire map area is the primary determinant of its size. These results make good sense: observations have shown that lower exponents are associated with higher-order goods and higher exponents are associated with lower-order goods; and that establishments offering higher-order goods (lower exponents) tend to cluster in the centre of a region—for example the central business district in a city—while establishments offering lower-order goods (higher exponents) tend to be widely dispersed and sensitive to local competition.

Forrester's models of urban and world dynamics fall naturally into the dynamic modelling category, though in their original form they did not explicitly model spatial process. The urban-dynamics model has been interpreted in a spatial context by Wils (1974). And the world model has been taken a step further by Mesarovic and Pestel (1975) who, recognizing that world regions, of which they considered ten, evolve at different speeds and have different resource depletion rates, were able to demonstrate that catastrophes, if they arise, are more localized than the fully aggregated model built by Forrester suggested.

8.3. Historical connections

Perhaps the biggest problem in systems analysis is to integrate system structure and function with system evolution. Gerard (1969) distinguished three facets of systems: being, or structure; behaving, or function; and becoming, which subsumes development and evolution. Gerard argued that history produces structure, which produces function, which produces history, and so forth, not circlewise, but spiralling up the steps of an evolutionary ladder. The relation between structure, function, and history has never been resolved. This may partly be because of a flaw in arguments like Gerards: structure, function, and history, rather than occurring one after the other, occur at the same time—they are all an expression of a system's dynamical behaviour. But how can structure, function, and history be treated simultaneously? This is still an unanswered question but recently some light has been shed on the matter. Prigogine (1978), discussing the behaviour of physico-chemical systems away from equilibrium, has shown that non-equilibrium can be a source of order. He offers the instance of a layer of water between two planes. If the difference in temperature between the two planes is large enough, the state of rest in the water becomes unstable and convection starts. In fact, small convection currents are always present in the water, even when at rest, and appear as fluctuations from the average state; but, below a critical temperature difference, the fluctuations are damped and disappear. On the other hand, above a critical temperature difference, certain fluctuations are amplified and give rise to large-scale currents which are stabilized by exchanges of energy with the outside world. By showing that laws of thermodynamics near to equilibrium, which are very general, differ from laws of thermodynamics away from equilibrium, which may be very specific, Prigogine (1978) demonstrated that the time evolution of thermodynamic systems may involve the crossing of successive thresholds from one state to another in a definite order—this is like wandering along certain paths in a phase space—and in this sense introduced history into physics and chemistry.

A not dissimilar sort of research has been carried out by Kirkby (1978) who investigated thresholds of many geomorphological processes. Kirkby found that within the phase space of some geomorphological systems, there commonly appear to be unstable and stable regions. The systems will tend to move towards, and manifest themselves within, stable states, of which there may be more than one. Examples Kirkby developed include a system in which soil depth influences the rate of weathering such that weathering is at is slowest in either very thin soils or very thick soils. The dynamics of this system determine that neighbouring sites which start at almost the same thickness will with time diverge to form exposed bedrock tors and stable soil layers: states in between the two are unstable and so are seldom found in the landscape. Other thresholds which can be explained in a similar manner are those between braiding and meandering and between smooth slopes and valleys.

Work of this kind, which reveals how, from the time of its formation, a system unfolds, is related to Thom's work on catastrophe theory, and, despite its being fashionable, looks like offering some answers to question of origin and evolution which have been puzzling geographers for many years.

8.4. A final observation

Many geographers judge systems analysis as unproductive or even pernicious. One wonders if this judgement is based on a careful study of the technique or on ignorance of it. What harm can come from using systems analysis in geography? Probably none. On the contrary, as this book has endeavoured to show, systems analysis is a useful implement which has already produced interesting results and turned up some unexpected findings. Surely, if geographers eschew the system-mongers, they will forgo in systems analysis a sophisticated and powerful method of study.

References

ABLER, R., ADAM, J. S., and GOULD, P. R. (1971), *Spatial organization: the geographer's view of the world*, Prentice-Hall, Englewood Cliffs.
ACKOFF, R. L. (1976), 'Towards a system of systems concepts', in *Systems behaviour*, 2nd edition (ed. J. Beishon and G. Peters), pp. 105–12, published for the Open University Press by Harper & Row, London, reprinted from *Management Science*, 17 (1971).
AHNERT, F. (1971), *A general and comprehensive theoretical model of slope profile development*, Final Report, Grant No. DA-ARO-D-31-124-G1020, Department of Geography, University of Maryland.
ALLEN, J. R. L. (1974), 'Reaction, relaxation and lag in natural sedimentary systems: general principles, examples and lessons, *Earth–Science Reviews*, **10,** 263–396.
ALFELD, L. E. and MEADOWS, D. L. (1974), 'A systems approach to urban revival', in *Readings in urban dynamics*: volume 1, (ed. N. J. Mass), pp. 41–56, Wright–Allen Press, Cambridge, Massachusetts.
ANDERSON, P. W. (1972), 'More is different', *Science*, **177,** 393–6.
ANUCHIN, V. A. (1973), 'Theory of geography', in *Directions in geography* (ed. R. J. Chorley), pp. 43–63, Methuen, London.
ASHBY, W. R. (1966), *Design for a brain*, Wiley, London.

BENNETT, R. J. (1975). 'The representation and identification of spatio-temporal systems: an example of population diffusion in North-West England', *Transactions of the Institute of British Geographers*, **66,** 73–94.
—— (1978), *Spatial time series: analysis, forecasting and control*, Pion Press, London.
——, and CHORLEY, R. J. (1978), *Environmental systems: philosophy, and control*, Methuen, London.
BERRY, B. J. L. (1959), 'Ribbon developments in the urban business pattern', *Annals of the Association of American Geographers*, **49,** 145–55.
—— (1973), 'A paradigm for modern geography', in *Directions in geography*, (ed. R. J. Chorley), pp. 3–21, Methuen, London.
——, BARNUM, H. G., and TENNANT, R. J. (1962), 'Retail location and consumer behaviour', *Regional Science Association, Papers and Proceedings*, **9,** 65–106.
BERTALANFFY, L. von (1971), *General system theory*, Penguin Books, Harmondsworth.
BLALOCK, H. M. (1964), *Casual inferences in non-experimental research*, University of North Carolina Press, Chapel Hill, North Carolina.
BLEDSOE, L. J., and VAN DYNE, G. M. (1971), 'A compartment model simulation of secondary succession', in *Systems analysis and simulation in ecology*, volume 1, (ed. B. C. Patten), pp. 479–511, Academic Press, New York and London.
BORMANN, F. H., LIKENS, G. E., and MELILLO, J. M. (1977), 'Nitrogen budget of an aggrading northern hardwood forest ecoystem', *Science*, **196,** 981–3.
BOULDING, K. E. (1956), 'General systems theory—the skeleton of science', *General Systems*, **1,** 11–17.
—— (1962), *A reconstruction of economics*, Science Editions, New York.
BRACEY, H. E. (1962), 'English central villages: identification. distribution and

functions', *Lund Studies in Geography, Series B, Human Geography*, **24,** 169–90.

Bradford, M. G., and Kent, W. A. (1977), *Human geography. Theories and their applications*, Oxford University Press, Oxford.

Briggs, L. I., and Pollack, H. N. (1967), 'Digital model of evaporite sedimentation', *Science*, **155,** 453–6.

Bunge, W. (1962), 'Theoretical geography', *Lund Studies in Geography, Series C, General and Mathematical Geography*, 1.

Cannon, W. B. (1932), *The wisdom of the body*, Norton, New York.

Chadwick, G. (1971), *A systems view of planning*, Pergamon Press, Oxford.

Champion, A. G. (1972), 'Urban densities in England and Wales: the significance of three factors', *Area*, **4,** 187–92.

Chapman, G. P. (1974), 'Perception and regulation: a case study of farmers in Bihar', *Transactions of the Institute of British Geographers*, **62,** 71–93.

Checkland, P. B. (1972), 'A systems map of the universe', in *Systems behaviour*, 1st edition, (ed. J. Beishon and G. Peters), pp. 50–55, published for the Open University Press by Harper & Row, London, reprinted form the *Journal of Systems Engineering*, **2,** (1971).

Chen, C. W., and Orlob, G. T. (1975), 'Ecologic simulation for aquatic environments', in *Systems analysis and simulation in ecology*, volume 3 (ed. B. C. Patten), pp. 475–588, Academic Press, New York and London.

Chenery, H. B., and Clark, P. G. (1959), *Interindustry economics*, Wiley, New York.

Chisholm, M. D. I. (1967), 'General systems theory and geography', *Transactions of the Institute British Geographers*, **42,** 42–52.

—— (1975), *Human geography: evolution or revolution?*, Penguin Books, Harmondsworth.

Chorley, R. J. (1966), 'The application of statistical methods to geomorphology', in *Essays in Geomorphology* (ed. G. H. Dury), pp. 275–387, Heineman, London.

—— (1973), 'Geography as human ecology', in *Directions in geography* (ed. R. J. Chorley), pp. 155–69, Methuen, London.

——, and Haggett, P. (1969), *Network analysis in geography*, Edward Arnold, London.

——, and Kennedy, B. A. (1971), *Physical geography. A systems approach.* Prentice-Hall International, London.

Clymer, A. B. (1972), 'Next-generation models in ecology', in *Systems analysis and simulation in ecology*, volume 2, (ed. B. C. Patten), pp. 533–69, Academic Press, New York and London.

Cooke, R. U. (1970), 'Morphometric analysis of pediments and associated landforms in the western Mohave desert, California', *American Journal of Science*, **269,** 26–38.

—— (1971), 'Systems and physical geography', *Area*, **3,** 212–16.

Cooper, C. F. (1969), 'Ecosystem models in watershed management', in *The ecosystem concept in natural resource management* (ed. G. M. Van Dyne), pp. 309–24, Academic Press, New York.

——, Blasing, T. J., Fritts, H. C., Oak Ridge Systems Ecology Group, Smith, F. M., Parton, W. J., Schreuder, G. F., Sollins, P., Stoner, W., and Zich, J. (1974), 'Simulation models of the effects of climatic change on natural ecosystems', *Proceedings of the Third Conference on the Climatic Assess-*

ment Program, pp. 550–62. US Department of Transportation, Transportation Systems Center, Cambridge, Massachusetts.

CRAWFORD, N. H., and LINSLEY, R. K. (1966), *The Stanford Watershed Model Mark IV, Technical Report* **39,** Department of Civil Engineering, Stanford University.

CURRY, L. (1962), 'Climatic change as a random series', *Annals of the Association of American Geographers,* **52,** 21–31.

CURSON, P. H. (1976), 'Household structure in nineteenth century Auckland', *New Zealand Geographer,* **32,** 177–93.

DI TORO, D. M., O'CONNOR, D. J., THOMANN, R. V., and MANCINI, J. L. (1975), 'Phytoplankton—zooplankton—nutrient interaction model for western Lake Erie', in *Systems analysis and simulation in ecology,* volume 3 (ed. B. C. Patten), pp. 423–74, Academic Press, New York and London.

DMOWSKI, R. M. (ed.) (1974), *Systems analysis and modelling approaches in environmental systems,* Polish Academy of Sciences, Warsaw.

EAGLESON, P. S. (1970), *Dynamic hydrology,* McGraw-Hill, New York.

EYRE, S. R. (1973), 'The spatial encumbrance', *Area,* **5,** 320–4.

FLEMING, G. (1970), 'Simulation of streamflow in Scotland', *Bulletin of the International Association of Scientific Hydrology,* **15,** 53–9.

FORRESTER, J. W. (1961), *Industrial dynamics,* MIT Press, Cambridge, Massachusetts.

—— (1969) *Urban dynamics,* The MIT Press, Cambridge, Massachusetts.

—— (1971), *World dynamics.* Wright–Allen Press, Cambridge, Massachusetts.

—— (1976), 'Understanding the counterintuitive behaviour of social systems', in *Systems behaviour,* 2nd edition (ed. J. Beishon and G. Peters), pp. 223–40, published for the Open University Press by Harper & Row, London.

GARNER, B. J. (1966), 'The internal structure of shopping centres', *Northwestern University, Studies in Geography,* **12.**

—— (1967), 'Models of urban geography and settlement location', in *Models in geography*', (ed. R. J. Chorley and P. Haggett), pp. 303–60, Methuen, London.

GERARD, R. W. (1957), :Units and concepts of biology', *Science,* **125,** 429–33.

—— (1969), 'Hierarchy, entitation and levels', in *Hierarchical structures* (ed. L. L. Whyte, A. G. Wilson, and D. Wilson), pp. 215–30, American Elsevier, New York.

GERSMEHL, P. J. (1976), 'An alternative biogeography', *Annals of the Association of American Geographers,* **66,** 223–41.

GOULD, P. R. (1969), *Spatial diffusion,* Association of American Geographers, Commission on College Geography, Resource Paper Series.

GOULD, S. J. (1966), 'Allometry and size in ontogeny and phylogeny', *Biological Review,* **41,** 587–640.

GRAY, J. (1976), 'Are marine base-line surveys worthwhile?', *New Scientist,* 29 Apr. 1976, pp. 219–21.

GREEN, H. L. (1955), 'Hinterland boundaries of New York city and Boston in southern New England', *Economic Geography,* **31,** 283–300.

GUNN, D. J. (1971), 'Computers and methods for computation', in *Chemical engineering,* volume 3 (ed. J. F. Richardson and D. G. Peacock), pp. 225–345, Pergamon Press, Oxford.

HÄGERSTRAND, T. (1967), 'On Monte Carlo simulation of diffusion', in *Quantitative geography*, Part I: *Economic and cultural topics* (ed. W. L. Garrison and D. F. Marble), pp. 1–32, Northwestern Studies in Geography, No. 13, Northwestern University Press, Evanston, Illinois.

HAGGETT, P. (1965), *Locational analysis in human geography* (first edition), Edward Arnold, London.

—— (1967), 'Network models in geography', in *Models in geography* (ed. R. J. Chorley), pp. 609–68, Methuen, London.

—— (1972), *Geography : a modern synthesis*, 1st edition, Harper & Row, New York and London.

——, CLIFF, A. D., and FREY, A. (1977), *Locational analysis in human geography* (Second edition; two volumes), Edward Arnold, London.

HALL, A. D., and FAGEN, R. E. (1956), 'Definition of system', *General Systems*, **1,** 18–28.

HALLAM, A. (1973), *A revolution in the earth sciences*, Oxford University Press, Oxford.

HAMILTON, H. R., GOLDSTONE, S. E., MILLIMAN J. W., PUGH, A. L. III., ROBERTS, E. B., and ZELLNER, A. (1969), *Systems simulation for regional analysis. An application to river-basin planning*, The MIT Press, Cambridge, Massachusetts.

HARRIS, B. (1972), 'Change and equilibrium in the urban system', in *Systems approach and the city*', (ed. M. D. Mesarovic and A. Reisman), pp. 68–86, North–Holland Publishing Company, Amsterdam and London.

HARRISON, H. L., LOUCKS, O. L. MITCHELL, J. W., PARKHURST, D. F., TRACY, C. R., WATTS, D. G., and YANNACONNE, J. C. Jr. (1970), 'System studies of D.D.T. transport', *Science*, **170,** 503–8.

HARRISON, W. (1968), 'Empirical equation for longshore current velocity', *Journal of Geophysical Research*, **73,** 6929–36.

HARVEY, D. W. (1969), *Explanation in geography*, Edward Arnold. London.

HARWELL, M. A., CROPPER, W. P., and RAGSDALE, H. L. (1977), 'Nutrient recycling and stability: a re-evaluation', *Ecology*, **58,** 660–6.

HEPPLE, L. W. (1974), 'The impact of stochastic process theory upon spatial analysis in human geography', *Progress in Geography*, **6,** 89–142.

HETT, J. M., and O'NEILL, R. V. (1974), 'Systems analysis of the Aleut ecosystem', *Arctic Anthropology*, **11,** 31–40.

HILL, D. M. (1965), 'A growth allocation model for the Boston region', *Journal of the American Institute of Planners*, **31,** 111–20.

HIRANO, M. (1975), 'Simulation of development process of interfluvial slopes with reference to graded form', *Journal of Geology*, **83,** 113–23.

HOOS, I. R. (1972), *Systems analysis in public policy: a critique*, University of California Press, California.

HORTON, R. E. (1945), 'Erosional development of streams and their drainage basins: hydrophysical approach to quantitative morphology, *Bulletin of the Geological Society of America*, **56,** 275–370.

HUFF, D. L. (1964), 'Defining and estimating a trade area', *Journal of Marketing*, **28,** 34–8.

HUGGETT, R. J. (1975), 'Soil landscape systems; a model of soil genesis', *Geoderma*, **13,** 1–22.

—— (1976), 'A schema for the science of geography, its systems, laws and models', *Area*, **8,** 25–30.

HUXLEY, J. S. (1932), *Problems of relative growth*, Methuen, London.

—— (1953), *Evolution in action*, Chatto & Windus, London.

ISARD, W., BRAMHALL, D. F., CARROTHERS, G. A. P., CUMBERLAND, J. H., MOSES, L. N., PRICE, D. O., and SCHOOLER, E. W. (1960), *Methods of regional analysis: an introduction to regional science*, The MIT Press, Cambridge, Massachusetts.

ISERMANN, R. (1975), 'Modelling and identification of dynamic processes—an extract', in *Modeling and simulation of water resources systems* (ed. G. C. Vansteenkiste), pp. 7–37. North-Holland Publishing Company, Amsterdam and Oxford.

JONES, E. (1964), *Human geography*, Chatto & Windus, London.

JORDAN, C. F., and KLINE, J. R. (1972), 'Mineral cycling: some basic concepts and their application in a tropical rain forest', *Annual Review of Ecology and Systematics*, **3,** 33–50.

——, and SASSCER, D. S. (1972), 'Relative stability of mineral cycles in forest ecosystems', *American Naturalist*, **106,** 237–53.

—— (1973), 'A simple model of strontium and manganese dynamics in a tropical rain forest', *Health Physics*, **24,** 477–89.

KEEBLE, D. E. (1971), 'Employment mobility in Britain', in *Spatial policy problems of the British economy* (ed. M. Chisholm and G. Manners), pp. 24–68, Cambridge University Press, Cambridge.

KIRKBY, M. J. (1971) 'Hillslope process-response models based on the continuity equation', in *Slopes: form and process* (ed. D. Brunsden), pp. 15–30, Institute of British Geographers, Special Publication No. 3, London.

—— (1978), 'The stream head as a significant geomorphic threshold', School of Geography, University of Leeds, *Working Paper* 216.

KLINE, J. R. (1973), 'Mathematical simulation of soil-plant relationships and soil genesis', *Soil Science*, **115,** 240–9.

KOESTLER, A. (1967), *The ghost in the machine*, Hutchinson, London.

—— (1969), 'Beyond atomism and holism—the concept of the holon', in *Beyond reductionism: new perspectives in the life sciences* (ed. A. Koestler and J. R. Smythies), pp. 192–232, Hutchinson, London.

KORELESKI, K. (1975), 'Types of soil degradation on loess near Kraków', *Journal of Soil Science*, **26,** 44–52.

KOWAL, N. E. (1971), 'A rationale for modelling dynamic ecological systems', in *Systems analysis and simulation in ecology*, volume 1, (ed. B. C. Patten), pp. 123–94, Academic Press, New York.

KRUMBEIN, W. C., and GRAYBILL, F. A. (1965), *An introduction to statistical models in geology*, McGraw-Hill, New York.

LAKSHMANAN, T. R., and HANSEN, W. G. (1965), 'A retail market potential model', *Journal of the American Institute of Planners*, **32,** 134–43.

LANE, E. W. (1957), 'A study of the shape of channels formed by natural streams in erodible materials', *U.S. Army Corp of Engineers*, Sediment Series, No. 9.

LANGTON, J. (1972), 'Potentialities and problems of a systems approach to the study of change in human geography', *Process in Geography*, **4,** 125–79.

LASZLO, E. (1972), *Introduction to systems philosophy*, Gordon & Breach, London.

LEONTIEF, W. W. (1966), *Input—output economics*, Oxford University Press, Oxford.

LEOPOLD, L. B., and LANGBEIN, W. B. (1962), *The concept of entropy in landscape evolution*, US Geological Survey, Professional Paper 500A.

——, WOLMAN, M. G., and MILLER, J. P. (1964). *Fluvial processes in geomorphology*, Freeman, San Francisco.

LESLIE, P. H. (1945), 'The use of matrices in certain population mathematics', *Biometrika*, **33,** 183–212.

—— (1948), 'Some further notes on the use of matrices in population mathematics', *Biometrika*, **35,** 213–45.

LEVISON, M., WARD, R. F., and WEBB, J. W. (1973), *The settlement of Polynesia: a computer simulation*, Minneapolis.

LEWIS, E. G. (1942), 'On the generation and growth of a population', *Sankhya*, **6,** 93–6.

LOTKA, A. J. (1925), *Elements of physical biology*, Williams & Wilkins, Baltimore (reprinted as *Elements of mathematical biology* in 1956 by Dover Publications New York.)

LOUCKS, O. L. (1977), 'Emergence of research on agro-ecosystems', *Annual Review of Ecology and Systematics*, **8,** 173–92.

LOWRY, I. S. (1964), *A model of metropolis*, Memorandum RM-4035-C, The RAND Corporation Santa Monica, California.

MABOGUNJE, A. L. (1970), 'Systems approach to a theory of rural–urban migration', *Geographical Analysis*, **2,** 1–18.

MARTIN, R. L., and OEPPEN, J. E. (1975), 'The identification of regional forecasting models using space—time correlation functions', *Transactions of the Institute of British Geographers*, **66,** 95–118.

MARUYAMA, M. (1960), 'Morphogenesis and morphostasis', *Methodos*, **12,** 251–96.

—— (1963), 'The second cybernetics: deviation–amplifying mutual causal processes', *American Scientist*, **51,** 164–79.

MASS, N. J. (ed.) (1974), *Readings in urban dynamics*, volume 1, Wright–Allen Press, Cambridge, Massachusetts.

MASSER, I. (1972), *Analytical models for urban and regional planning*, David & Charles, Newton Abbott.

MAYR, E. (1970), *Population, species and evolution*, The Belknap Press of Harvard University Press, Cambridge, Massachusetts.

MCPHERSON, H. J. (1969), 'Flow, channel and floodplain characteristics of the lower Red Deer River, Alberta', in *Geomorphology. Selected Readings* (ed. J. G. Nelson and M. J. Chambers), pp. 257–79, Methuen, Toronto.

MCQUITTY, L. L. (1957), 'Elementary linkage analysis for isolating orthogonal and oblique types and typal relevancies', *Educational and Psychological Measurement*, **17,** 207–29.

MEADOWS, D. H., MEADOWS, D. L., RANDERS, J., and BEHRENS, W. W. III (1972), *The limits to growth. A report for the Club of Rome's project on the predicament of mankind*, A Potomac Associates book published by Universe Books, New York; also published in 1974 by Pan Books, London.

MELTON, M. A. (1957), *An analysis of the relations among elements of climatic, surface properties and geomorphology*, Office of Naval Research Project NR 389–042, Technical Report 11. Department of Geography, Columbia University, New York.

—— (1959), 'A derivation of Strahler's channel-ordering system', *Journal of Geology*, **67,** 345–6.

MESAROVIC, M. D., and PESTEL, E. (1975), *Mankind at the turning point: the second report to the Club of Rome*, Hutchinson, London.

——, and TAKAHARA, Y. (1975), *General systems theory: mathematical foundations*, Academic Press, New York and London (volume 113 in Mathematics in Science and Engineering, edited by R. Bellman).

MILLER, J. G. (1965), 'Living systems', *Behavioral Science*, **10,** 193–237 and 337–411.

MILNER, C. (1972), 'The use of computer simulation in conservation management', in *Mathematical models in ecology* (ed. J. N. R. Jeffers), pp. 249–75, Blackwell Scientific Publications, Oxford.

MORISAWA, M. E. (1964), 'Development of drainage systems on an upraised lake floor', *American Journal of Science*, **262,** 340–54.

MORRILL, R. L. (1965), 'The Negro ghetto: problems and alternatives', *Geographical Review*, **55,** 339–61.

MUNTON, R. J. C. (1969), 'The economic geography of agriculture', in *Trends in geography* (ed. R. U. Cooke and J. H. Johnson), pp. 143–52. Pergamon Press, Oxford.

NACE, R. L. (1969), 'World water inventory and control', in *Water, earth and man* (ed. R. J. Chorley), pp. 31–42, Methuen, London.

NARROLL, R. S., and BERTALANFFY, L. von (1956), 'The principle of allometry in biology and the social sciences', *General Systems*, **1,** 76–89.

NOBLE, J. V. (1974), 'Geographic and temporal development of plaques', *Nature*, **250,** 726–9.

NORDBECK, S. (1971), 'Urban allometric growth', *Geografiska Annaler*, **53B,** 54–67.

NORTH, G. R. (1975), 'Theory of energy-balance climate models', *Journal of the Atmospheric Sciences*, **32,** 2033–43.

ODUM, H. T. (1971), *Environment, power, and society*, Wiley–Interscience, London.

—— (1976), 'Macroscopic minimodels of man and nature', in *Systems analysis and simulation in ecology*, volume 4 (ed. B. C. Patten), pp. 249–80, Academic Press, New York and London.

O'NEILL, R. V. (1975), 'Modeling in the eastern deciduous forest biome', in *Systems analysis and simulation in ecology*, volume 3, (ed. B. C. Patten), pp. 49–72, Academic Press, New York and London.

——, and BURKE, O. W. (1971), *A simple systems model for DDT and DDE movement in the human food-chain*, Eastern Deciduous Forest Biome ORNL-IBP-71-9, Oak Ridge National Laboratory, Oak Ridge, Tennessee.

PARK, R. A. *et al.* (1975), 'A generalized model for simulating lake ecosystems', *Simulation* (August, 1974), pp. 33–50.

PATTEN, B. C. (1971) 'A primer for ecological modeling and simulation with analog and digital computers', in *Systems analysis and simulation in ecology*, volume 1 (ed. B. C. Patten), pp. 3–121, Academic Press, New York and London.

——, and WITKAMP, M. (1967), 'Systems analysis of 134-caesium kinetics in terrestrial microsms', *Ecology*, **48,** 813–24.

——, BOSSERMAN, R. W., FINN, J. T., and GALE, W. G. (1976), 'Propagation of cause in ecosystems', in *Systems analysis and simulation in ecology*, volume 4, (ed. B. C. Patten), pp. 457–579, Academic Press, New York and London.

PIANKA, E. R. (1974), *Evolutionary ecology*, Harper & Row, New York.
PLATT, J. (1969a), 'What we must do', *Science*, **166,** 1115–21.
—— (1969b), 'Theorems on boundaries in hierarchical systems', in *Hierarchical Structures*, (ed. L. L. Whyte, A. G. Wilson and D. Wilson), pp. 201–13. American Elsevier, New York.
POLLARD, D. (1978), 'An investigation of the astronomical theory of the ice ages using a simple climate-ice sheet model', *Nature*, **272,** 233–5.
POPPER, K. (1974), 'Scientific reduction and the essential imcompleteness of all science', in *Studies in the philosophy of biology. Reduction and related problems*, (ed. F. J. Ayala and T. Dobzhansky), pp. 259–84, Macmillan, London.
PRIGOGINE, I. (1978), 'Time, structure, and fluctuations', *Science*, **201,** 777–85.

RAINES, G. E., BLOOM, S. G., and LEVIS, A. A. (1969), 'Ecological models applied to radionuclide transfer in tropical ecosystems', *Bioscience*, **19,** 1086–91.
RAY, D. M., VILLENEUVE, P. Y., and ROBERGE, R. A. (1974), 'Functional prerequisites, spatial diffusion and allometric growth', *Economic Geography*, **50,** 341–51.
RAYNER, J. N. (1972), *Conservation, equilibrium, and feedback applied to atmospheric and fluvial processes*, Association of American Geographers, Commission on College Geography, Resource Paper No. 15.
REES, P. H., and WILSON, A. G. (1977), *Spatial demographic analysis*, Edward Arnold, London.
REICHLE, D. E., O'NEILL, R. V., KAYE, S. V., SOLLINS, P., and BOOTH, R. S. (1973), 'Systems analysis as applied to modelling ecological processes', *Oikos*, **24,** 337–43.
REIF, B. (1973), *Models in urban and regional planning*, Leonard Hill Books, Aylesbury.
REILLY, W. J. (1931), *The law of retail gravitation*, G. P. Putnam & Sons, New York.
RIDLEY, J. C., and SHEPS, M. C. (1966), 'An analytic simulation model of human reproduction with demographic and biological components', *Population Studies*, **19,** 297–310.
RODIN, L. E., and BASILEVICH, N. I. (1967), *Production and mineral cycling in terrestrial vegetation*, Translated from the Russian, Oliver & Boyd, London.
ROGERS, A. (1968), *Matrix analysis of interregional population growth and distribution*, University of California Press, Berkeley, California.
ROSS, P. J., HENZELL, E. F., and ROSS, D. R. (1972), 'The effects of nitrogen and light in grass-legume pastures—a systems analysis approach', *Journal of Applied Ecology*, **9,** 535–56.
ROWE, J. S. (1961), 'The level-of-integration concept and ecology', *Ecology*, **42,** 420–7.
RUHE, R. V., and WALKER, P. H. (1968), 'Hillslope models and soil formation. I. Open systems', *Transactions of the Ninth International Congress of Soil Science*, Adelaide, **4,** 551–60.
RYKIEL, E. J., and KUENZEL, N. T. (1971), 'Analog computer models of "The Wolves of Isle Royale" ', in *Systems analysis and simulation in ecology*, volume 1, (ed. B. C. Patten), pp. 513–41, Academic Press, New York and London.

SALERNO, J. (1973), 'Sensitivity in the world dynamics model', *Nature*, **244,** 488–92.

Sasscer, D. C., Jordan, C. F., and Kline, J. R. (1971), 'A mathematical model of tritiated and stable water movement in an old-field ecosystem', in *Radionuclides in ecosystems. Proceedings of the Third National Symposium on Radioecology*, pp. 915–23, CONF-710501-P1, US Atomic Energy Commission.

Scheidegger, A. (1970), *Theoretical geomorphology*, second, revised edition. Springer-Verlag, Berlin.

Schlager, K. J. (1965), 'A land use plan design model', *Journal of the American Institute of Planners*, **31,** 103–10.

Schultz, A. M. (1969), 'A study of an ecosystem: the arctic tundra', in *The ecosystem concept in natural resource management* (ed. G. M. Van Dyne), pp. 77–93, Acedimic Press, New York and London.

Sellers, W. D. (1969), 'A global climatic model based in the energy balance of the earth-atmosphere system', *Journal of Applied Meteorology*, **8,** 392–400.

—— (1973), 'A new global climatic model', *Journal of Applied Meteorology*, **12,** 241–54.

Shah, M. M. (1975), 'Systems engineering approach to agricultural and rural development systems', in *Modeling and simulation of water resources systems* (ed. G. C. Vansteenkiste), pp. 565–77, North-Holland Publishing Company, Amsterdam and Oxford.

Shreve, R. L. (1966), 'Statistical law of stream numbers', *Journal of Geology*, **74,** 17–37.

Simon, H. A. (1954), 'Spurious correlations: a causal interpretation', *Journal of the American Statistical Association*, **49,** 467–79.

—— (1962), 'The architecture of complexity', *Proceedings of the American Philosophical Society*, **106,** 467–82.

Smailes, A. E. (1971), 'Urban systems', *Transactions of the Institute of British Geographers*, **53,** 1–14.

Smith, J. M. (1974), *Models in ecology*, Cambridge University Press, London.

Spanner, D. C. (1964), *Introduction to thermodynamics*, Academic Press, London.

Spedding, C. R. W. (1975), 'The study of agricultural systems', in *Study of agricultural systems* (ed. G. E. Dalton), pp. 3–19, Applied Science, London.

Speight, J. G. (1974), 'A parametric approach to landform regions', in *Progress in geomorphology. Papers in honour of David L. Linton* (ed. E. H. Brown and R. S. Waters), pp. 213–230, Institute of British Geographers Special Publication No. 7, London.

Stoddart, D. R. (1967), 'Organism and ecosystem as geographical models', in *Models in geography* (ed. R. J. Chorley and P. Haggett), pp. 511–48, Methuen, London.

—— (1969), 'World erosion and sedimentation', in *Water, earth and man* (ed. R. J. Chorley), pp. 43–64, Methuen, London.

Strahler, A. N. (1952), 'Dynamic basis of geomorphology', *Bulletin of the Geological Society of America*, **63,** 923–38.

Sugden, D., and Hamilton, P. (1971), 'Scale, systems and regional geography', *Area*, **3,** 139–44.

Sumner, G. N. (1978), *Mathematics for physical geographers*, Edward Arnold, London.

Taaffe, E. J., Morrill, R. L., and Gould, P. R. (1963), 'Transport expansion in underdeveloped countries: a comparative analysis', *Geographical Review*, **53,** 503–29.

TAYLOR, P. J. (1977), *Quantitative methods in geography*, Houghton Miffin, Boston.

TERJUNG, W. H. (1976), 'Climatology for geographers', *Annals of the Association of American Geographers*, **66,** 199–222.

THOM, R. (1972), *Stabilité structurelle et morphogénèse*, Benjamin, New York.

THOMAS, E. (1965), *A structure of geography: a proto-unit for secondary schools*, High School Geography Project, Association of American Geographers, University of Colorado, Boulder, Colorado.

THOMAS, R. W. (1975), 'Some functional characteristics of British city central areas: an application of allometric principles', *Regional Studies*, **9,** 369–78.

—— (1977), *An introduction to quadrat analysis*, Concepts and techniques in modern geography (CATMOG) No. 12, Study Group in Quantitative Methods, Institute of British Geographers, London.

THOMPSON, D. W. (1917), *On growth and form*, Cambridge University Press, London.

TOMOVIĆ, R. (1963), *Sensitivity analysis of dynamic systems*, McGraw–Hill New York.

TRUDGILL, S. T. (1977), *Soil and vegetation systems*, Oxford University Press, Oxford.

USHER, M. B. (1972), 'Developments in the Leslie matrix model', in *Mathematical models in ecology* (ed. J. N. R. Jeffers), pp. 29–60, Blackwell Scientific Publications, Oxford.

VOLTERRA, V. (1926), 'Fluctuations in the abundance of a species considered mathematically', *Nature*, **118,** 558–60.

WADDINGTON, C. H. (1957), *The strategy of the genes: a discussion of some aspects of theoretical biology*, Macmillan, London.

—— (1968). 'The basic ideas of biology', in *Towards a theoretical biology*, 1. *Prolegomena*. (ed. C. H. Waddington), pp. 1–32, University of Edinburgh Press, Edinburgh.

—— (1977), *Tools for thought*, Paladin, St. Albans.

WAGGONER, P. E., and STEPHENS, G. R. (1970), 'Transition probabilities for a forest', *Nature*, **225,** 1160–1.

WARNTZ, W. (1973), 'New geography as general spatial systems theory—old social physics writ large?', in *Directions in geography*, (ed. R. J. Chorley), pp. 89–126, Methuen, London.

WATSON, R. A. (1969), 'Explanation and prediction in geology', *Journal of Geology*, **77,** 488–94.

WATT, A. S. (1947), 'Process and pattern in the plant community', *Journal of Ecology*, **35,** 1–22.

WATT, K. E. F., YOUNG, J. W., MITCHINER, J. L., and BREWER, J. W. (1975), 'A simulation of the use of energy and land at the national level', *Simulation* (May 1975), pp. 129–53.

WHITE, R. W. (1977), 'Dynamic central place theory: results of a simulation approach', *Geographical Analysis*, **9,** 226–43.

WHITTLESEY, D. (1954), 'The regional concept and the regional method', in *America geography: inventory and prospect* (ed. P. James and C. F. Jones), pp. 19–68, Syracuse University Press, Syracuse (for the Association of American Geographers).

WIEGERT, R. G. (1975), 'Simulation models of ecosystems', *Annual Review of Ecology and Systematics*, **6,** 311–38.
WILLIAMSON, M. H. (1972), *The analysis of biological populations*, Edward Arnold, London.
—— (1967), 'Introducing students to the concepts of population dynamics', in *The teaching of ecology* (ed. J. M. Lambert), pp. 169–76, Symposium of the British Ecological Society No. 7.
WILS, W. (1974), 'Metropolitan population growth, land area, and *urban dynamics* model', in *Readings in urban dynamics*: volume 1, (ed. N. J. Mass), pp. 103–20. Wright–Allen Press, Cambridge, Massachusetts.
WILSON, A. G. (1972), 'Theoretical geography: some speculations', *Transactions of the Institute of British Geographers*, **57,** 31–44.
—— (1974), *Urban and regional models in geography and planning*, John Wiley & Sons, London.
——, and KIRKBY, M. J. (1975), *Mathematics for geographers and planners*, Oxford University Press, Oxford.
WOOD, M. J., and SUTHERLAND, A. J. (1970), 'Evaluation of digital catchment model on New Zealand catchments', *Journal of Hydrology* (*New Zealand*), **9,** 323–35.
WOOD, S. R. (1975), 'A catchment simulation model developed for urban and urbanising catchments with particular reference to the use of automatic optimisation techniques', in *Modeling and simulation of water resources systems* (ed. G. C. Vansteenkiste), pp. 209–18, North-Holland Publishing Company, Amsterdam and Oxford.

YEATES, M. H. (1974), *An introduction to quantitative analysis in human geography*, McGraw-Hill, New York.
YOST, C., and NANEY, J. W. (1975), 'Earth-dam seepage and related land and water problems', *Journal of Soil Water Conservation*, **30,** 87–91.

ZIPF, G. K. (1949), *Human behaviour and the principle of least effort*, Cambridge University Press, Cambridge.

Index